AF443223

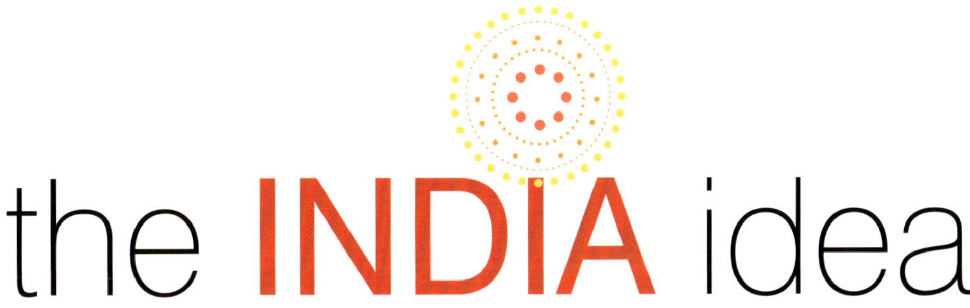

Concept: Navdeep Suri
Creatives: Shobit Arya
Captions: Shobit Arya, Nandini Gupta. Factual information provided by respective photographers.
Copy Editing: Nandini Gupta, Jyoti Kumari
Design & Layout: Supriya Saran, Gautam Kumar
Review: Anand Khandelwal, Swati Chopra, Jyothi Rajamohan

© Wisdom Tree 2011

First published in 2011 by:
Wisdom Tree,
4779/23, Ansari Road,
Darya Ganj, New Delhi-2
Ph.: 23247966/67/68
wisdomtreebooks@gmail.com

The views and opinions expressed in this book are those of the individual contributors and do not necessarily reflect the views of the publishing house.

ISBN 978-81-8328-251-2

Printed in India

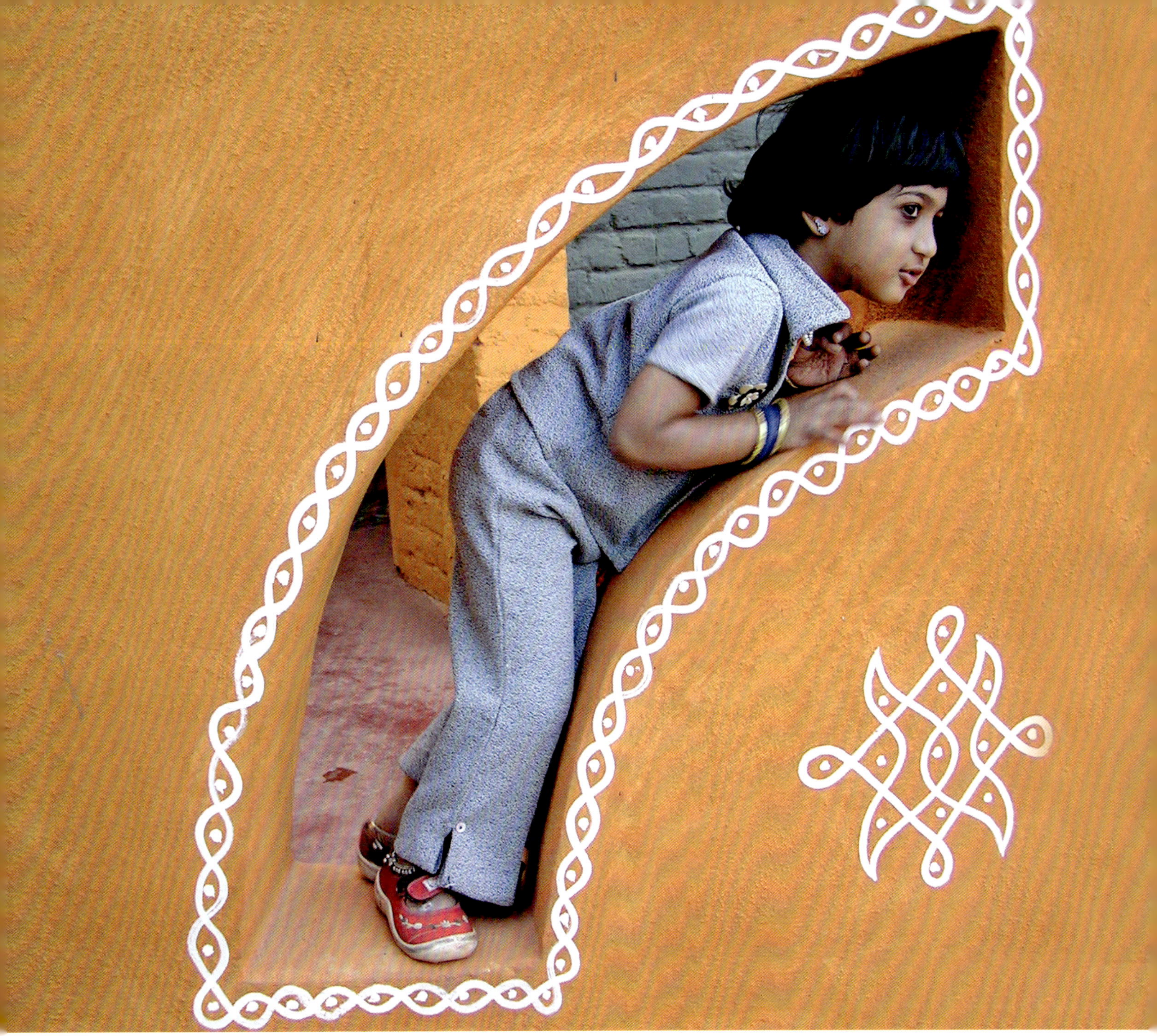

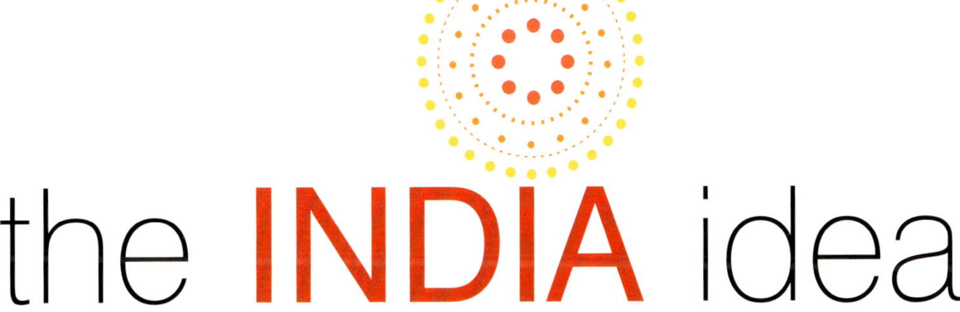

the INDIA idea

Edited by
LK Sharma

Photo Research & Editing by
Shobit Arya

Contents

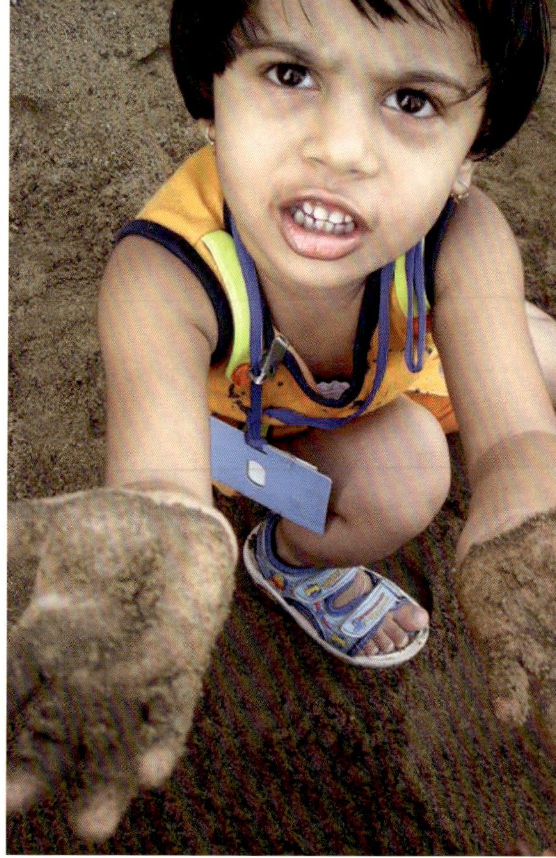

Creating the INDIA idea

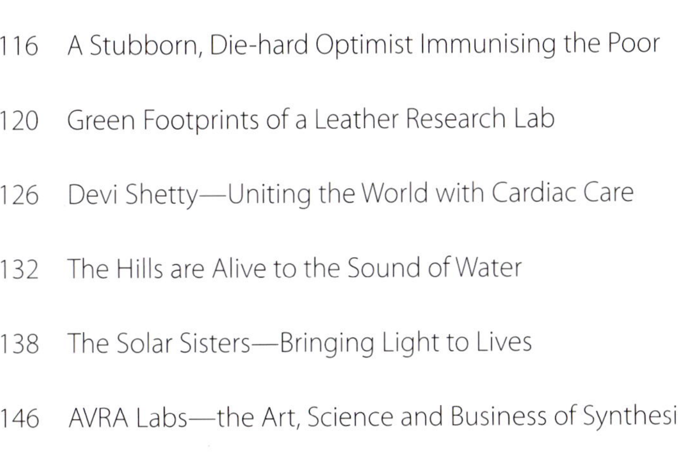

Sharing the INDIA idea

Introduction

LK Sharma

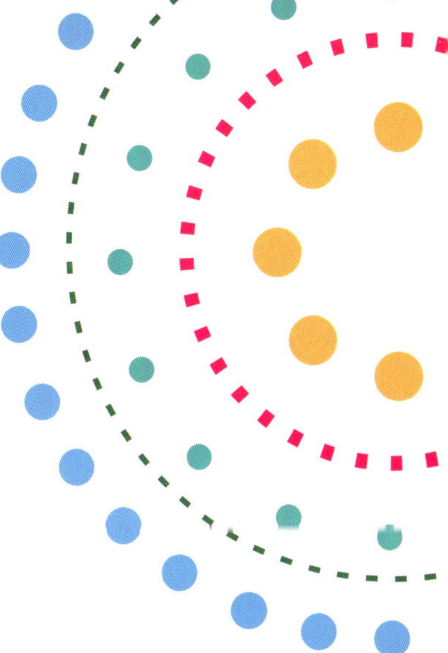

Change has speeded up. The interplay between society and technology has intensified, each influencing the other. Its full force is being felt in India. Satellites and mobile phones are providing education, expert advice, weather information and other critical information to farmers and fishermen, patients and students. The Indian villager once had to trudge miles to sell her farm produce and then was forced to sell it at whatever price she got. Today she gets the *mandi* prices on the cell phone before setting out for the market, averting exploitation. On a Village Resource Centre (VRC) computer, the farmer sees an image of his small plot taken by a remote sensing satellite, gets to know its soil characteristics and a list of recommended crops.

Innovations in technology, governance and business model are the key drivers of change. The combined use of space technology and mobile telephony has unleashed a revolution. These provide services in the fields of health, education, agriculture, fisheries, natural resources management, urban planning and disaster relief. A villager's critical heart condition can be remotely diagnosed by a super specialist practising in a distant city. Banks of brick and mortar that failed to reach the thousands of villages now enable unlettered customers to operate bank accounts with the help of cards and hand-held biometric swiping machines that capture their thumb impressions.

These are sophisticated technologies being used for societal applications. But many Indians do not wait for a 'technology giver' to solve their problems or improve their lives. A Punjab villager sitting in his farm pieces together a rugged multi-utility vehicle to tap the full potential of his water pump engine discarded motor parts and bits of scrap. He drives his home-made multi-purpose vehicle, enjoying the curious glances of onlookers. He shows playfulness with the material and with the language in naming his vehicle—Maruta. The name is derived from the better-known Maruti car! This kind of creative improvisation has matured to the extent that a standard English dictionary will soon have to include the Hindi word *Jugaad*.

Examples of *Jugaad* as also of systemic innovations, reflecting creativity at every level, are proliferating. These come through channels as diverse as the sophisticated space, nuclear and biotechnology programmes, corporate research laboratories and semi-literate farmers and street-side mechanics doing their own problem-solving research. The individual effort based on intuition, experience and limited resources is known to have paid off in thousands of cases. These are 'Indian Solutions'. Many of them are found to be extremely useful in other developing countries. These then become 'Solutions *from* India'—transforming lives not only in India but also in other parts of the world.

All developing countries face somewhat similar problems, including the resource crunch that makes expensive solutions unusable. That is why official aid agencies, India's Technical Cooperation Mission as well as international charitable organisations and activists track such innovations in this country. Villagers in Afghanistan get to use for the first time a public convenience that has been built by an Indian NGO using its special technique. An AIDS patient in Ivory Coast can afford to buy a drug for his very survival because India-made drugs are inexpensive. A disabled Iraqi is able to walk, thanks to the Jaipur Foot. When solar power lights up homes in 162 villages in twenty-eight African countries, an Indian solution becomes a solution *from* India!

Experts have attributed the rise of India, as of some other countries in Asia, to the strengths in low-cost manufacturing and knowledge-based industries and the competence in innovating new technologies and business models. India has become a big laboratory for testing ideas. There are problems needing solutions that are best perceived by the Indian eyes. Thus an innovative inverter business comes up when an Indian entrepreneur surveys the scene and finds that the western multinationals considered inverters as outdoor aids while India needed power backup systems for day-to-day use!

The global companies are learning from India lessons of simplicity and affordability, what Dr RA Mashelkar calls Gandhian engineering. They are looking towards India for ideas for products that can be used by millions of potential customers with limited purchasing power. Once such a made-for-India product is launched, these companies discover that it has a market even in some developed countries affected by economic down-turn. An American gets the benefit of a relatively inexpensive electrocardiogram machine originally developed for rural India. Customers in Europe get to use a people's car made in India. Siemens in India, which plans to cut product development costs by as much as 70 per cent, is working on forty-two products designed for Indian and other emerging markets. These include a solar-powered X-ray machine and foetal heart monitor. Bosch India developed a fuel injection system or the pump for diesel engines with the latest and stricter emission norms. This had an impact on the global diesel engine industry.

India's core competence in science and technology, as reflected in major projects, is now globally appreciated. How has that core competence impacted the lives of the people? This is what this book is about. The experience of gathering contents for this volume was akin to searching for a precious stone and stumbling on a large cache of jewels. India is not generating 'disruptive' technologies as yet but the small and big innovations and novel ways of using established technologies are having a 'disruptive' impact. This endeavour involves the government's scientific establishment, educational institutions, private companies, grass-roots

innovators and activists. Both the formal and informal sectors are demonstrating new techniques. When the two sectors join hands, one sees synergy's multiplier effect. A micro entrepreneur on bicycle carries mobile phone from person to person in villages, providing service to those who cannot afford even a cell phone. A bank correspondent goes with a hand-held swiping machine to deliver money to the account holders in remote areas who have yet to see a real bank. The induction of technology is improving governance, increasing transparency in the functioning of the authorities at every level and giving ordinary people access to useful information and improved techniques. An innovative use of Information Technology (IT) has helped eradicate illiteracy in the targeted communities. The expected and unforeseen gains of the explosive growth of mobile telephony are seen all around. It has empowered the weaker sections as nothing else was able to do before.

In the organised sector, the Indian drugs industry has made critical treatment affordable, winning for India the title of the 'Pharmacy of the World'. Surgeons and physicians turned social entrepreneurs have developed innovative health delivery business models to serve the people who could not have afforded the normal costs of diagnosis and treatment. Technical innovations are coupled with innovations in management.

The constant talk of Nano is not just about a car. The word has become a qualifying epithet being applied to any innovative product that reaches the unreached. Once you plan for 'inclusive development' you can't leave out the prime enabler—'Inclusive innovation'. Once you start thinking of the problems at the grass-roots, you have to think of an inexpensive water purifier to protect them against water-borne diseases. If you make a product affordable, a new and large clientele is ready to buy it. Hence those raising the slogan of 'inclusive development' are not just charities but hard-headed business leaders. The corporates have internalised the mantra given by the late management guru Prof CK Prahlad who drew their attention to the treasure lying at the 'Bottom of the Pyramid'. To illustrate his thesis, he collected some examples of Indian innovative products that enriched companies while enabling the resource-poor families to enjoy the fruits of development. Businesses have come to realise that doing good offers rewards in the present world itself. As an ITC executive who ushered in the company's e-Choupal initiative for farmers points out, their approach is to create, 'Fortune at the "Bottom of the Pyramid" for their shareholders through creation of fortune for the "Bottom of the Pyramid"'. This has resulted in a race to reach the unreached with innovative products and services.

Much more is yet to come. Growing technological connectivity, IT's open source movement and crowdsourcing-based Research and Development

A robust communication model across the spectrum, leading to a culture of knowledge sharing. (left) Farmers getting online information in an e-choupal in Satrik, Barabanki, Uttar Pradesh. Launched a decade ago by the Indian corporate giant, ITC, e-Choupal has brought the virtual world of the market at the doorsteps of the farming community in India, thereby ensuring that the farmers get maximum benefit from the sale of their produce. (below) Youngsters attending a session of TED India conference held at the Infosys campus in Mysore. TED stands for Technology, Entertainment and Design, and the first conference was held in 1984 in California, USA. India is one of the countries where TED has organised a conference that focuses on people's innovations.

(R&D) harness diffused creative knowledge. With its core competence in the field of IT, India is going to be a major contributor and beneficiary of the open source software movement. This model, now reaching out to other sectors, opens up immense new possibilities. With the Council of Scientific & Industrial Research (CSIR) initiating a project that may lead to the development of a TB drug through crowdsourcing, the trend will gather momentum. This again is a move towards inclusive innovation. These developments are good for India's ecology of innovations and for the resource-constrained end users of new products and services.

The ratio of the R&D that gets translated into marketable products is increasing. Dead ideas and prototypes pile up in the 'Valley of Death' but the mortality rate has been reduced. Even CSIR tries to ensure that the fruits of its scientific endeavour reach industry and society. A new culture in the official scientific institutions and the involvement of the NGOs and profit-making businesses ensure efficient marketing of the new products as well.

S&T inputs are being used for rural development by young entrepreneurs. Some of them move to villages when they get tired of making millions and being called techies. They return from Silicon Valley or exit the Indian Institute of Technology (IIT) hostels to apply their skills to the problems faced by the 'real' India. Unlike their fathers and uncles, these young men and women are not stricken by economic insecurity and thus they don't just feel for the poor but do something for them. They do not fear of failure. These activists have generated new useful solutions and configured imaginative projects around available solutions. A few pioneers were there in the sixties and seventies but now there are large numbers working in the areas of health, education, agriculture extension, energy and water conservation.

Societal and economic benefits accrue only when new techniques are coupled with innovations in project management and service delivery models. The stories in this book contain another significant lesson. Simple solutions based on an innovative

approach and combining traditional wisdom with modern knowledge, work if community participation is secured. That alone makes projects sustainable. An official agency at times due to inadequate understanding of the community picks up an inappropriate technology and then neglects community mobilisation. In the implementation of innovative projects, the NGOs and a unique government agency such as the Indian Space Research Organisation (ISRO) have shown the way. They provide a constant 'last mile link' with the beneficiaries. The community must first accept a novel project, participate in its execution and then maintain it. Many activists live among the villagers, gain their confidence and ensure that the beneficiaries have a stake in running the project.

The young breed of technologist-social activist, committed to rural development, is following the country's unique tradition in its scientific endeavour. India's eminent scientists from the pre-Independence days did not work in sealed ivory towers. To an extent, they were driven by social concerns. In some cases, such concerns even determined their choice of discipline or made them return to India. Dr MS Swaminathan chose to study the unfashionable agricultural science and then went on to do not just farm research but strategic planning to take the new knowledge to the farmers. He mobilised farmers through field visits. The Green Revolution involved the most significant innovation with societal applications. What Dr Swaminathan did in the field of agriculture, Dr Verghese Kurien did in the field of animal husbandry and was named the Father of the White Revolution. He is popularly called the Milkman of India. Noted cosmic ray physicist Dr Vikram Sarabhai, while formulating the space science and technology plan, gave considerable thought to societal applications and to innovative management of the hi-tech projects. This book features examples of hi-tech being used for rural development.

Another seam from which innovations are being mined actually lies at the 'Bottom of the Pyramid'. Here the faceless humble semi-literate 'innovators' learn by experience, tinker with their tools and come up with new ideas and products. The National Innovation Foundation (NIF) tracks such material through extensive field trips led by a noted management professor Anil Gupta. Thanks to his efforts, such innovators have gained visibility and recognition for their contributions. This has

improved the chances of their ideas being picked and translated into products. The view that ordinary people are also blessed with creativity is gaining acceptance. This also means fewer of their ideas perishing. The growing tribe of grass-roots innovators now gets respect from the elite hi-tech community. This is because prestigious global companies have started soliciting ideas from ordinary people, amateur enthusiasts, users of their products and potential customers. In-house experts are not the sole repository of knowledge. It is a salute to the creativity of the common man. This book gives illustrative examples of grass-roots innovations.

Ecology conducive to innovation is being created by the ongoing radical economic and social transformation. The process of globalisation coupled with economic liberalisation has influenced manufacturing, marketing and industrial R&D. The large scientific programmes have provided spin-off benefits, touching lives. The common man has tasted the fruits of IT and mobile telephony. While economic inequity has grown, efforts to bridge the digital divide have yielded results in selected pockets. The inherent character of some new technologies makes it possible for these to reach the disempowered and resource-poor. These technologies have become powerful weapons in the armoury of development planners, social entrepreneurs and activists. Paradoxically, these happen to be more sophisticated but less expensive! They are spreading fast because India has the advantage of being a late starter and thus less burdened by the legacy systems. The first ring tone that millions of Indians heard was on a mobile phone!

The World Wide Web, an expanding infrastructure, new business opportunities, growing competition, official promotional schemes and the arrival of venture capitalists—all are playing a part. A critical contributory factor is the emergence of an aspirational and self-confident India. In this India, the people are striving to live better, work better. They are unwilling to live with problems that now seem amenable to solutions. As they compete, succeed and prosper, they look for new ways of doing things.

India is witnessing an innovations movement. Big corporates, village craftspersons, housewives, entrepreneurs, IIT students, roadside mechanics, government scientists, the NRIs back from California—all are involved in this new grand enterprise. It has no single leader. Dr RA Mashelkar started campaigning for such a movement on becoming the head of the CSIR in 1995 but perhaps even he had not imagined that it would gather such a momentum within such short span of time.

—LK Sharma is a senior journalist
and the text editor of this book.

Indian Needs, Indian Solutions

How did thousands of villagers get to see a moving image for the first time? Not because of the 100-year-old Indian film industry but thanks to the pioneering Satellite Instructional Television Experiment (SITE) of the seventies. India was one of the first two users of this direct broadcast technology with societal applications. Today, the VRCs established by ISRO provide information in the local languages, meeting the multiple needs of the communities.

The Computer-based Functional Literacy Programme initiated by Tata Consultancy Services uses a mix of specially designed software, multimedia presentations, and printed materials to teach an uneducated person to read in about forty hours spread across ten to twelve weeks. It does not need trained teachers. The emphasis is on words, rather than letters, and the process is styled to suit the learner. The students acquire a 300-500 word vocabulary in their own language and dialect, enough for everyday requirements, such as reading destination signs on buses, simple documents and newspapers.

Bhoomi, an e-governance project, has brought about a sea change in maintaining and administering land records in Karnataka. The online management and delivery of land records being transparent protects the farmers from possible exploitation. They access the data base and obtain a printed copy of their land deeds at the computerised land record kiosks in taluk offices, for a fee of ₹15.

Since, the second computer screen faces the client; they can also see the transaction being performed.

E-Choupal, a network of virtual commodity market has demonstrated how a private company could establish a mutually profitable relationship with over 4 million farmers who have their produce to sell and other items to buy. The heart of an e-Choupal is a computer with internet facility that establishes a two-way communication channel between the company ITC and the villagers. The computer is set up in the house of a progressive farmer who helps others access prices of their crops in the *mandi* and the price that company running the e-Choupals is paying. This empowers the farmers and they do transactions armed with correct information and cannot be misled by any intermediaries. This is done through a web portal in the local language. There are now mobile e-Choupals also.

As a student, Santosh Ostwal had watched his eighty-two-year-old grandfather getting up at midnight to hobble along for a mile to switch on the water pump at his orange farm because power was not available during the day. After getting an engineering degree, he devised a system that triggers the irrigation pump remotely through a mobile phone.

Mobile telephone messages are used by young men and women to send signals of love, by advertisers to sell their wares and by activists to

unleash an SMS uprising. But they are also sent to save lives. Veerampattinam, a fishing village near Puducherry, used to lose seven to eight people every year to the tidal waves. But since 1998 not a single life has been lost at sea. Now the fishermen know in advance when it is dangerous to go fishing.

Several people are killed by fires resulting from the cooking gas leakage. Gautam Kumar aged twenty-six, along with a friend, developed a fire alarm system, Suraksha that sends mobile telephone messages to five registered users. The device consists of a sensor that smells gas and then sends the SMS. Kumar won the 2011 Technology Review Social Innovator award for this groundbreaking innovation.

It is not just human lives but the lives of sheep and goats are also being saved through mobile messages. A Pune-based NGO transmits messages in Marathi about simple herbal medicines and vaccines to the nomadic shepherds of Maharashtra. The NGO has provided them compact solar chargers cum lamps. In two hours the mobiles are charged and then the charger doubles as a lamp.

A generic pharma company, in collaboration with UNICEF, produced the first of its kind Mother-Baby Pack that extends world-class anti-retrovirals to the infected expectant mothers and plays a pivotal role in the prevention of mother-to-child transmission of HIV/AIDS.

A new meningitis vaccine made by an Indian company is being used in Africa. It is designed to be affordable and hence as many as 450 million people in the African meningitis belt will be inoculated by 2015.

Indigenous knowledge has helped develop a new drug and also improve the livelihood of the Kani tribe. The scientists of the Kerala-based Tropical Botanic Garden and Research Institute (TBGRI) used the tribal know-how of the wild plant, Arogyapaacha's properties to develop an anti-stress and anti-fatigue drug named Jeevani.

Traditional knowledge was used by a small entrepreneur to challenge mighty multinationals and build the largest hair-wash products company, CavinKare. He had very little resources but had ideas and made his products win a large share of the market through cost-cutting R&D and an innovative marketing strategy. The time-tested natural products that were used to formulate shampoo included *reetha* nuts, *shikakai*, almonds, hibiscus leaves and some herbs. The concept of small sachets, as much as its quality and low prices popularised his shampoo.

About 10 million people in India are afflicted with cerebral palsy, autism, mental retardation, and aphasia. Ajit Narayanan aged twenty-nine, developed a device, AVAZ, to help those suffering from speech disorders. It converts limited head or finger movements into speech. AVAZ falls in the

category of Augmentative and Alternative Communication technologies. Narayanan won the *Technology Review*'s Innovator of the Year Award for 2011. AVAZ is innovative because it can generate speech in Indian languages. And, of course, it costs one-tenth of the price of similar devices. A typical Indian innovator seems to first set a price at which the product must be made available.

What is common between Delhi's police, the Indonesian President's security force, a Japanese auto parts manufacturer and an American wine company? They are all using the 'nonClonableID' technology developed by a Pune-based company. This nanotechnology-based system provides fool proof anti-counterfeiting solutions with applications in the pharma, agrochemical and currency sectors, most vulnerable to counterfeiting. The 'nonClonableID' technology-based identity cards for the Delhi Police are highly protected, compact and cannot be duplicated by anybody, including the manufacturers. A portable electronic reader can authenticate a card at any place.

Pest resistant and high oil-yielding varieties of mint plant developed by CSIR are being cultivated on more than 4,00,000 hectares of land. These have been adopted by 20,000 farmers and have generated 40,000,000 man-days of employment.

To help brick-makers an NGO developed TARA BrickMek machine that produces uniform high quality clay bricks through an efficient system that integrates multiple functions, is easy to operate and cuts down coal consumption.

Four young women of Gujarat, just out of college, went to a drought-prone area and generated one of the earliest success stories of innovation and activism in the field of water. In the Bhal area, water tables were falling and salinity was rising and they were inspired by a Dalit woman who had to walk for miles everyday to get drinking water for the family. They consulted engineers and implemented a rainwater harvesting scheme by lining surface tanks with plastic so that these retained rain water. The Indian Petrochemicals Corporation Limited (IPCL) pitched in with help. The results surprised the officials used to the big piped water system. Such initiatives have provided relief to those living in the areas afflicted with water-borne diseases and saline or undrinkable water.

In Rajasthan, Tarun Bharat Sangh, headed by Magsaysay Award winner Rajendra Singh, has popularised rain water harvesting by reviving dried up water ponds and small rivers. The rejuvenation of the traditional water harvesting structures started a movement that has spread to other areas. Rajendra Singh is known as the 'Jal Purush' or the 'Water Man of Rajasthan.'

Inclusive Innovation:
Getting More from Less for More

RA Mashelkar

The next century will belong to India, which will become a unique intellectual and economic power to reckon with, recapturing all its glory, which it had in the millennia gone by.

Innovation today is widely recognised as a major source of competitiveness and economic growth for advanced and emerging economies alike. Its significant role in creating jobs, generating incomes and improving living standards is now well understood. However, it is yet to be fully recognised as a powerful tool for promoting 'inclusive growth'. Truly inclusive innovation will have to cater to the needs of 4 billion people, whose income levels are less than US$2 per day. This calls for a paradigm shift. Getting more (performance) by using less (resources) for more (profit) is what every enterprise aims for. But we can create 'inclusive growth' only when we achieve more (performance) by using less (resources) for more (people). Let us call it MLM, 'more from less for more people' paradigm.

The objective of MLM type of innovation would not be just to produce low performance, cheap knock-off versions of rich country technologies so that they can be marketed to the poor. Rather, the objective is to harness sophisticated science and technology to invent, design, produce and distribute high performance technologies at prices that can be afforded by a majority of the people. The MLM challenge involves inclusive innovation that creates extremely affordable products as well as offers extremely affordable services through business process innovations, workflow innovations and so on. Innovation is all about doing things differently to make a big difference.

If the MLM paradigm is made into a reality then the poor can potentially have the same 'functional and emotional experience' as the rich. For example, a two-wheeler scooter owner in India today can afford to buy the world's cheapest car, Tata Nano, costing US$2,000. And this is a proper car. It is comfortable, fuel efficient and environment-friendly. This is an example of an innovation that is 'inclusive'. The challenge of serving these billions of people lies in moving from 'low-cost' to 'ultra low-cost' and from 'incremental innovation' to 'disruptive innovation'. Let us illustrate this challenge further. To how many of the following questions can we answer in 'yes'?

- Can we make a Hepatitis-B vaccine priced at US$20 per dose available at a price that is 40 times less?

- Can we make an artificial foot priced at US$10,000 available at a price that is 300 times less?
- Can we make high-quality cataract eye surgery available, not at US$3,000, but at a price that is 100 times less?
- Can we make a mobile phone call at not 8 cents per minute but at a price that is 8 times lower?

Incredible as it may sound; all such MLM targets have been met in India. Here are some brilliant examples of such MLM innovations:

MLM through Product Innovation: Tata Nano

The idea of Tata Nano was conceptualised by Ratan Tata himself. He challenged young designers of Tata Motors to design and develop a very low-cost four-wheeler without compromising on the performance. The price of US$2,000 was emphasised in all departments, namely design, development, production, materials, logistics and sales. The Tata Nano team drew ideas from different sources, from helicopters to two-wheelers. For example, the mechanism of helicopter seats was used to get a solution for the design of Tata Nano seats. Window winding mechanism was inspired by helicopter windows. The dashboard, fuel lines and lamps drew inspiration from two-wheelers. Thus they produced a small car with contemporary styling, spacious interiors and high standards of performance. Tata Nano has a rear mounted 624 cc, 35 bhp engine, a maximum speed of 125 km/hr with a fuel consumption of 23 km/litre. It meets the Euro IV emission standards. The low-cost assembly line of Tata Nano, its innovative partnership with component manufacturing and the innovative business model for automobile dealerships have set new benchmarks for the global automobile industry.

'Nano' Refrigerator

Tata Nano was produced taking into account the low paying capacity of a large section of the people. But India has many other constraints. It has the world's largest population deprived of electricity. Effective refrigeration in rural areas can help people extend their access to food and also essential drugs. Godrej and Boyce took up this challenge. They launched ChotuKool, world's cheapest refrigerator with a US$69 price tag. The portable,

top-opening unit weighs only 7.8 kg, uses high-end insulation to stay cool for hours without power and consumes half the energy used by regular refrigerators. To achieve its efficiency, the ChotuKool does not use a compressor; instead it runs on a cooling chip and a fan similar to that used in computers. Like computers, it can run on batteries. In a true MLM spirit, it has only twenty parts as opposed to more than 200 parts in a normal refrigerator. The ChotuKool was designed in consultation with village women to ensure its acceptability. It is distributed by members of a micro-finance group. The villagers act as marketers and earn a commission of approximately US$3 per fridge sold. This fridge is targeted at households, who earn approximately US$5 a day, of whom there are almost 100 million in India. A true MLM story!

The public sector too is generating products that fall in the MLM category. A US$30 laptop for students has been designed by scientists of India's premier science and technology institutions like IIT and Indian Institute of Science (IISc). The low-cost laptops are proposed to be distributed in millions in schools by the Government of India.

MLM through Business Process Innovation

The telecom industry revolution in India, specifically in wireless communication, is another example of MLM, with brilliant business process innovation. This industry now adds around 20 million subscribers per month. The cost of a minute of a mobile phone time is less than 1 cent, the lowest in the world. The cost of one SMS text message has dropped dramatically! This journey began with an audacious dream of a visionary leader Dhirubhai Ambani, who challenged his team to innovate and bring down the cost of a phone call to that of a postcard in India. The Reliance team did not follow the traditional model of purchasing telecom equipment on a cost-per-subscriber basis, which meant paying a massive upfront cost per subscriber fee to vendors. Reliance paid them for the volume of traffic of voice that flowed through the equipment per unit time. Reliance also

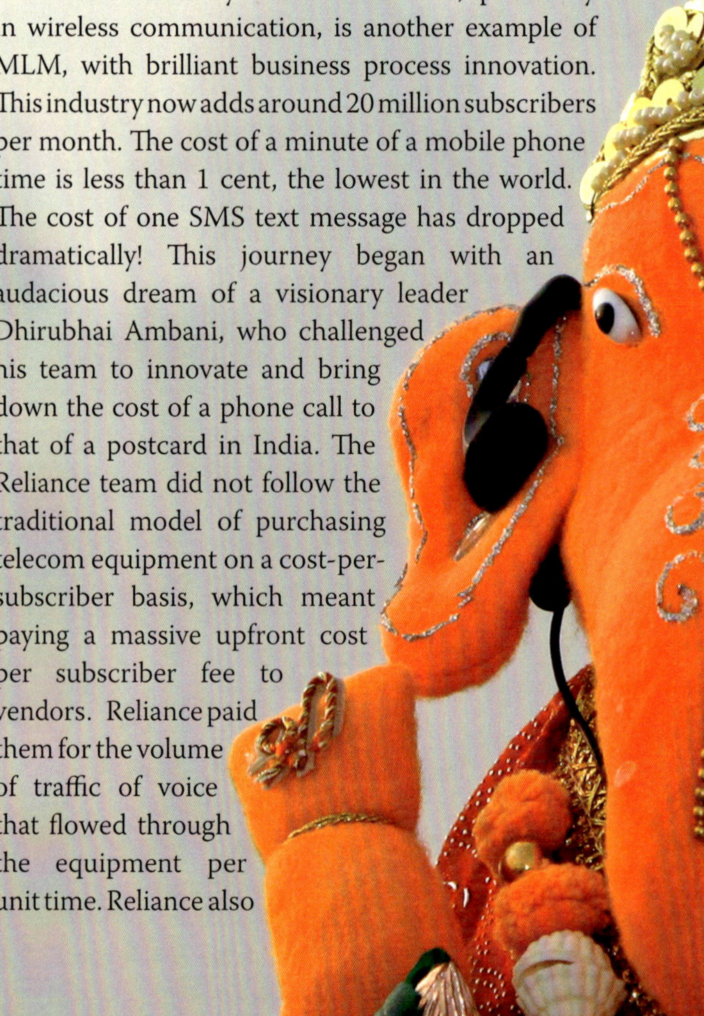

The past merges with the present. A toy model of the god of auspicious beginnings, Ganesha, listening to an MP3 player against the backdrop of the famous Taj Mahal.

pioneered some groundbreaking marketing strategies including free text messages, free phones, free incoming calls, and more. With Reliance's entry, the call rates dropped exponentially. It's deal with equipment suppliers set the benchmark for the lowest equipment prices in the world.

The remarkable success of Airtel, another mobile telephone operator, highlighted the importance of the business model innovation. Airtel shifted the focus from Average Revenue Per User (ARPU) to contribution per minute and from vertical integration to outsourcing. It sought the best partners in the world and with each vendor; it negotiated win-win agreements that focused on growth. Further, capital expenditures (fixed costs) were converted into operating expenses (variable costs). It built rapidly an ecosystem of value-added application developers and a large distribution channel piggy backing on existing small Indian retailers. It also built a system of contract governance that provided for flexibility, learning and change.

MLM through Work Flow Innovation

Aravind Eye Care System was started by late Dr G Venkataswamy with a mission to eliminate 'needless blindness'. The cost of a typical cataract surgery in the US is around US$3,000. Aravind has managed to bring down the cost to US$30, performing around 3,00,000 surgeries per year. Aravind Eye Care has developed a cost-effective revenue model so that thousands of blind poor persons can be operated on for free or nearly free. But can such cheap eye surgery deliver quality results? A comparison of the data on some post-surgery parameters shows that Aravind Eye Care outperforms the Royal College of Ophthalmic Surgeries in UK. So, Aravind Eye Care's innovation is not about getting 'less from less'. It is about getting 'more from less'. And that too for 'more and more people'.

MLM by Involving Masses

Since MLM innovations are for masses, an interesting question is this. Can we get the masses involved in doing the MLM innovation? The answer is in the affirmative.

India's NIF set up in the year 2000 recognised that India just does not have 1 billion mouths to feed; it has 1 billion minds that can think! NIF started scouting, promoting and rewarding the grass-roots innovations and innovators. Thousands of such innovations have been documented and encouraged.

To take just one typical example, NIF identified an innovation on a rural washing machine that can work without electricity because it is pedal-driven. A Kerala school girl, Remya, developed it because she had this incredible combination of constraints coupled with her aspirations. Her father was down with cancer. Her mother was perennially ill. She had to change three buses to go to school. She had to come home, wash her clothes and do her studies. She created this washing machine, so that she could read, while the clothes were being washed, while she pedalled away.

Does MLM make business sense? MLM products and services will no longer be motivated by the concern about fulfilling the obligation of corporate social responsibility by the enterprises. MLM products and services are emerging as perhaps the biggest business opportunity of the coming decade. Most of the growth in consumer spending is expected to come from people in emerging markets, who have a much lower spending capacity than traditional middle class consumers in developed countries. This market can be served only by MLM products and services. Indian companies are particularly well positioned to take advantage of this opportunity. They have direct access to the world's second largest emerging market, one in which a huge low-income group is poised to enter the middle class. An explosion of consumer demand, spread across a range of low and middle-income segments, will allow Indian businesses to experiment with different scaling strategies, making the cost of pursuing MLM business models much lower for firms in India than for many competitors in other emerging markets.

If MLM business innovation models have to thrive, and in turn drive accelerated inclusive growth what kind of leadership qualities will be required? We have identified at least five.

First, MLM CEOs must develop a deep commitment to inclusive growth, which will force them to think of unserved customers, be they rural poor, who don't have access to telephones or urban poor, who don't get emergency medical services. Companies often start by asking, 'Given our cost structure, which segments can we serve?' They should ask, 'Given that we need to cater to the unserved, what should our cost structure be?'

Second, MLM CEOs must have clear vision with a human dimension, for example, helping poor Indians travel safely and affordably with their families; using connectivity to improve people's work and lives; and enabling patients to buy cheap medicines.

Third, MLM CEOs must establish ambitious goals and clear time frames for achieving them. Companies should ask, 'What is our on-the-moon project?' Or, as they do in India's boardrooms, 'What is our Nano project?'

Fourth, MLM leaders must force project teams to work within self-imposed boundaries that stem from a deep understanding of consumers. That will result in a novel, outside-in view of innovation. The language inside their organisations should be about consumers as people, suppliers as partners, and employees as innovators.

And finally, MLM CEOs must continuously ask, 'What if we change the way we operate to reduce costs and focus on return on capital employed, not just on operating margins? If we reduce prices enough and make our products available to the poor, won't there be explosive growth as they quickly find uses for and buy our offerings?'

Inclusive innovation through the MLM strategy forces us to measure opportunity by the 'ends' of innovation—what people actually get to enjoy—as opposed to just an increase in their means. This rationale invokes a return to the traditional case for innovation—its ability to produce breakthrough improvements in the quality of life—alongside the usual objective of increased competitiveness.

The combination of constraints and aspiration can provide an explosive trigger for extreme and disruptive innovation. The MLM way of innovation, anchored on the solid foundation of affordability and sustainability, will help us design a sustainable future for the mankind.

—RA Mashelkar is a Bhatnagar Fellow,
National Chemical Laboratory.

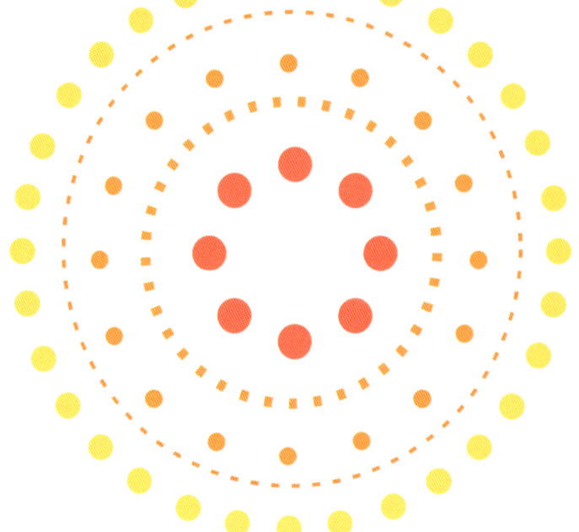

The Decade of Innovations—
a Blueprint

Sam Pitroda

Nations across the world are facing challenges in the areas of health, agriculture, education, environment, energy and governance which demand more efficient and sustainable solutions. These solutions should meet the needs of the many, especially those at the 'Bottom of the Pyramid', to promote a more inclusive society. Innovation can play a vital role in this process because innovations meet the needs which cannot be met by conventional products, processes and institutional forms. These yield significant social and economic gains. Innovations have the potential to redefine and reshape everything—from products and services, to governance, organisations, processes, people, economy, institutions, business and technology.

Innovations have been a part of Indian heritage and tradition—from the discovery of zero and the decimal system to being the cradle of an ancient civilisation and major world religions, from being home to the first global universities such as Nalanda and Takshshila to architectural and engineering marvels. In the recent times, India made successful strides in agriculture, milk productivity, space, atomic energy and telecom. However, this spirit of innovation has somehow been lost or subsumed in the development discourse of our nation. An innovation culture has to be triggered again in order to respond to India's unique needs and challenges in the competitive global marketplace.

We need to create a new model of 'Inclusive Innovation' for India which can provide solutions for the people at the 'Bottom of the Pyramid'. India needs more 'frugal innovation' that produces more 'frugal cost' products and services that are affordable by people at low levels of incomes without compromising the safety, efficiency and utility of the products. These innovations should not have an adverse impact on the environment to be sustainable in the long-term. This model can lead the way in solving the challenges of development, demography and disparity for developing economies the world over.

India's newly established National Innovation Council (NIC) has been mandated to implement a strategy for inclusive innovation and prepare a Roadmap for Innovation 2010-2020. It is planned to have a system that boosts innovation performance in the country. The cost-effective, executable and out-of-the box plan will drive innovation in the country by focusing on these key parameters:

- Platform: Widespread innovations in products, services, processes and across verticals to collectively create a strong and robust innovation society. Innovation should be widely distributed over the

Vittala Temple, Hampi, Karnataka. Digital Hampi, a unique collaborative project conceptualised by Ramachandra Budihal uses a software to resurrect the ancient glory of the ruins of Hampi. Visitors can wear 3-D glasses fitted with a camera. Following their hand gestures, the computer starts recreating the lost world in virtual reality. Budihal first demonstrated this project at the TED India conference in Mysore in 2009.

whole spectrum of economic activity, that is, across sectors (not just hi-tech), and types of innovations (not just formal R&D projects).

The broader platform for innovations will cover products, services, organisations and institutions, processes, R&D, science and technology, governance, social and cultural milieu and mindset.

- Inclusion: The council will focus on using innovation as a tool to eliminate disparity and meet the needs of the many in the best possible manner. The aim will be to create a model of inclusive innovation that underlines access, affordability and quality, and fosters innovations at the grass-roots and delivers 'more' value for 'many'.

  Such innovation will have features such as awareness, access, affordability, availability, scalability, sustainability, quality, pervasive growth, and innovations for and by the people at the 'Bottom of the Pyramid'.

- Ecosystem: A strong innovative ecosystem is critical for creating an innovation society. An innovative ecosystem must facilitate the birth of new ideas and also provide platforms for the successful implementation of these ideas. It is a dynamic system shaped by the interactions among multiple players such as the government, firms, schools and education and research institutions, financial institutions, individual innovators, customers and users, NGOs and the media.

  The council will facilitate the creation of necessary ecosystems through incentives and awards, innovation clusters at universities, innovative business clusters, innovation in the medium and small sectors of industry, organisational autonomy and flexibility, policies and programmes, new institutions and infrastructure, risk venture capital, intellectual property patents and the Information and Communication Technology (ICT) tools.

- Drivers: Such an innovation strategy has to be driven by some key ideas and goals. The council will focus on the key drivers such as

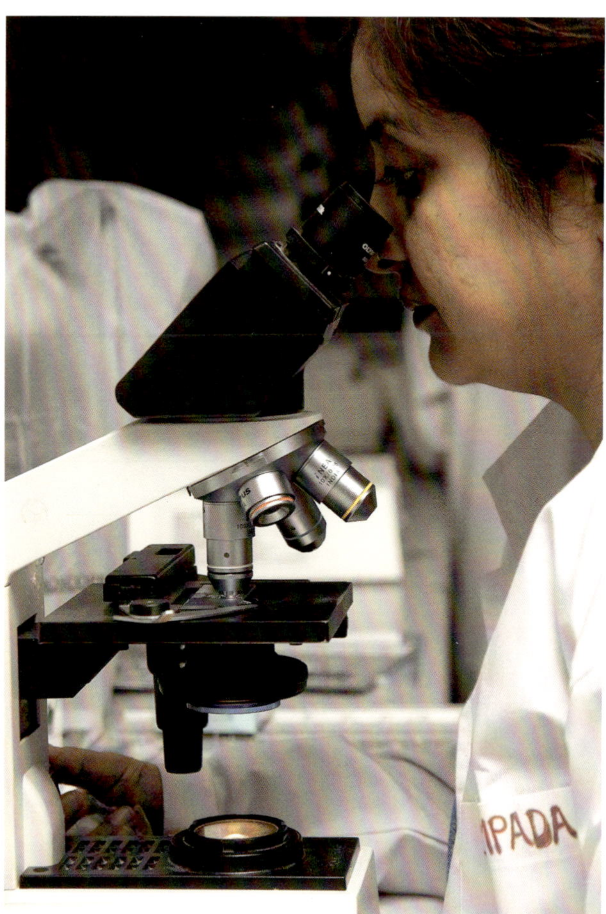

multidisciplinary approach, collaborative, transformative, generational change vs incremental change, durable vs disposable, need vs demand, nature vs nurture, locally relevant, globally connected and competitive, focus on the leading edge.

- Wider Discourse: The innovation discourse will include alternative dialogue, which often creates bypasses in the system to improve the current way of doing things. The aim is to have many divergent voices, views, mode of doing things to impact the end result qualitatively and quantitatively. The room for divergent discourse is especially critical in the official and organisational processes. The council will expand space for discourse on innovation through all available means.

This five-pronged plan will foster innovations by democratising information; identifying and empowering domain experts at the national, state and district levels. This will ensure institutional autonomy, freedom, flexibility, accountability and transparency. Community participation and the consultative process will be aimed at improving planning and governance.

The council will act as a catalyst for stimulating the innovation ecosystem to develop solutions for the country's agenda of inclusive growth. The formal institutional structure to promote innovations will include state innovation councils as well as councils for the key sectors. The sectoral bodies will plan to increase the nation's core competence in different fields with a view to make India more competitive. An Inclusive Innovation Fund will seek to plug the venture capital gap in the system. Innovation clusters will form regional hubs of innovation. Innovation centres in universities will bring the academic community and industries closer. A National Innovation Portal will be designed to galvanise resources on innovation. Of course, an effective outreach strategy will seek to bring about changes in the mindset. Through an institutional mechanism, international collaborations will be promoted and global thinking and experience leveraged.

Innovations require not just inputs and capacity but also a reformed political economy. The council will create a constituency for innovation, with enthusiastic participation by the government, academia, industry and the people. An innovation ecosystem should be able to generate hi-tech products as well as enhance the quality of life for everybody. With this intention, the President of India has declared the current decade as the 'Decade of Innovations'.

—Sam Pitroda is Adviser to the Prime Minister of India (Public Information Infrastructure and Innovations) and Head, National Innovation Council.

A scientist at work at a lab in Mumbai. Every year, medical colleges and universities across the country produce some 4,000 scientists and 1,000 PhDs. Thanks to these hardworking young individuals, India is fast developing into an R&D hub.

(facing page top) A Buddhist monk at the ruins of Nalanda in Bihar. Founded in the fifth century AD, Nalanda was the first residential international university of the world, with a strength of 10,000 students who were taught by 2,000 teachers. A massive, modern Nalanda International University is being set up in the state, with an aim of reviving the ancient learning tradition associated with the place.

(facing page bottom) A remarkable feat of creativity, Hawa Mahal in Jaipur. The pink sandstone, five-storeyed structure with latticed windows, was built by Sawai Pratap Singh in 1799. The ruler added another marvel to the glorious architecture of the city, well planned by the famous astronomer-king, Sawai Jai Singh II, in the early half of the eighteenth century.

creating
the INDIA idea

From **Green** to an Ever-green Revolution

MS Swaminathan

India's Green Revolution was the result of an innovative project that integrated research, education, and an extension service and then mobilised the peasants for technological upgradation of farm practices. This breakthrough provided food security to a country that was once forced to live from 'ship-to-mouth'. Later it also posed a challenge of how not to let it degenerate into a 'greed revolution'.

In the India of the sixties, the policy makers felt concerned that despite the development of the agricultural infrastructure, food production failed to keep pace with population increase. Various projects designed to increase farm productivity had yielded very little result. India had to import as much as 10 million tonnes of wheat in 1966. It was observed that traditional wheat and rice varieties, if fed with mineral fertilisers, tended to lodge because of their tall height and their straw. Since wheat and rice plants need about 25 and 20 kilograms of nitrogen respectively to yield 1 tonne of grain, it became clear that low yields would continue unless the architecture of the plant can be redesigned to enable it to utilise higher quantities of nutrients and irrigation water. It was necessary to breed the high-yielding varieties that can take advantage of good soil fertility management and irrigation water availability. A research strategy was developed for this purpose. Innovation rather than mere expansion of the farm area became a critical element of the new plan.

At that stage, Dr Orville Vogel of the Washington State University, Pulman, USA, announced the release of the winter wheat variety Gaines, capable of yielding over 10 tonnes per hectare. That variety was based on the Norin dwarfing gene identified by Dr Gonziro Inazuka at the Norin Experiment Station in Japan. I wrote to Dr Vogel in 1960 requesting for seeds of Gaines. He sent the seeds but mentioned that winter wheat, which required long days and mild temperature to flower and set grain, will not do well under the short day conditions prevailing during the rabi season (November to April) in North India. He suggested that I should obtain seeds of spring wheat varieties with the same dwarfing gene from Dr Norman E Borlaug. This eminent scientist was busy developing semi-dwarf varieties under a Rockefeller Foundation-Mexican Government programme in Mexico using the Norin dwarfing genes made available to him by Dr Vogel.

I wrote to Dr Borlaug requesting seeds of a wide range of semi-dwarf material. He promptly replied saying

(page 30) Women at the langar kitchen at the Ulsoor Gurdwara, the largest Sikh shrine in Bengaluru. Guru Nanak Dev, the first Sikh guru, laid the foundation of this free community meal service—putting aside all differences of gender, caste, colour, social status, age and religion—truly an innovative India idea.

(facing page) Kerala is known worldwide for its spices and cash crops. A pilot project to find out if cash crops can be grown in the salt-rich soil of India's coastal areas has been launched. If successful, it will transform agricultural production in coastal areas, which are becoming increasingly salty across the world.

that he will be happy to provide the material but a visit to the wheat-growing areas of India would help him to select it. Dr Borlaug visited us in March 1963. In September 1963, he sent seeds of the varieties that had already performed well, in addition to a wide range of segregating material. A five year plan (1963-68) was developed to prepare a road map for the Wheat Revolution.

The imported seeds were divided into several lots in order to get data simultaneously from Delhi, Ludhiana, Pant Nagar, Kanpur and Pusa during the rabi season of 1963-64. This helped us to study the genotype X environment interaction. It was clear that with the semi-dwarf varieties, we can treble the yield immediately (the average yield of wheat at that time was less than 1 tonne per hectare). In addition, the semi-dwarf wheat strains from Mexico showed wide adaptation to diverse growing conditions. A National Demonstration Programme was initiated in small farmers' fields during rabi season of 1964-65. We selected the fields of small farmers since yield results from demonstrations in rich farmers' fields will be attributed to their resources and not to technology. The small farmers chosen under this programme adopted the right agronomic practices and harvested on an average over 4 tonnes per hectare during April 1965. This revolutionary change in yield stirred up great enthusiasm among wheat farmers and there was a huge demand for seeds of the semi-dwarf high-yielding wheat varieties. More seeds were imported for rabi sowing in September 1966 and were distributed among all major wheat-growing states. At the same time, a Seed Village Programme was started in Jounti in Delhi State, where all farmers agreed to multiply the seeds of the semi-dwarf wheat varieties. Nearly a million hectares came under semi-dwarf varieties in the rabi season of 1967-68, resulting in an increase of nearly 7 million tonnes in wheat production in April 1968. Meanwhile, we also developed semi-dwarf varieties, such as Kalyan Sona and Sonalika, with amber grains and good chapatti-making qualities. The wheat harvest of 1968 marked the beginning of a new era in Indian agriculture. The then Prime Minister of India, Indira Gandhi released a special stamp to celebrate this success.

As to the development of new rice varieties, the work at the Indian Agricultural Research Institute (IARI) finally resulted in strains known as Pusa Basmati which has outstanding Basmati kind of

grain quality. After the official recognition of Pusa 1121 as a Basmati variety, the price of Pusa Basmati 4 went up to nearly ₹3,200 per quintal. Farmers in Haryana alone have derived a profit of over ₹2,000 crores from this variety. This illustrated the power of modern science to improve the economic well-being of farmers.

In the wake of this dramatic rise in productivity, I visited several farms in Punjab in 1967, and witnessed some undesirable farm practices that got associated with the Green Revolution. I warned the farmers against applying too much fertilisers and pumping out too much groundwater. Such practices, I said, would turn the Green Revolution into a 'greed revolution'. I cautioned against such a disaster in my address to the Indian Science Congress in Varanasi in January 1968. This led me later to coin the term 'Ever-green Revolution' to emphasise the need for improving productivity in perpetuity without associated ecological harm. The pathways to 'Ever-green Revolution' could be either organic farming or green agriculture.

New agriculture technologies like genomics and IT together with improved agronomic management should form the cornerstone of increasing agriculture productivity and profitability of small farms both in irrigated and rain-fed areas as well as in problem soils and coastal areas. Recombinant DNA technology has already resulted in the breeding of crop varieties possessing tolerance to salinity and drought as well as to serious biotic stresses caused by the triple alliance of pests, pathogens and weeds.

In this context, it is worth mentioning that Punjab Agricultural University has developed eco-technologies like sowing of wheat bed while saving 20 to 25 per cent water, leaf colour chart saving 15 per cent nitrogen application in rice, tensiometer-based irrigation scheduling, zero tillage technology for wheat and Integrated Pest Management (IPM) in cotton saving 40 per cent pesticides. Thus, there is vast scope both to promote Green Agriculture and to reduce the cost of production through enhanced factor productivity.

(top) A Kashmiri fruit-seller rows a shikara on the famous Dal Lake in Srinagar. The food processing industry in India is emerging as one of the major drivers of economic growth and will go a long way in helping the country emerge as the fruit basket of the world.

Kuttanad in Kerala is famous for rice cultivation. The farmers here practise below sea-level farming techniques, in which they rotate the cultivation of paddy or fish to get maximum yields. This way, they keep the weeds, pests and diseases in check, and make the best of both the monsoons and the post-monsoon weather conditions.

We are on the threshold of a biotechnology-IT revolution. Biotechnology does not imply only Genetically Modified Organisms (GMOs). There are many non-GMO applications, such as tissue culture for multiplying elite germplasm, bio-fertilisers, bio-pesticides and bio-remediation of groundwater as well as marker-assisted breeding. In the case of GMOs, safe and responsible use should be ensured. Organic farming permits the use of varieties developed by marker-assisted breeding.

The third technological revolution relevant to agriculture is the eco-technology movement. This involves the appropriate integration of frontier sciences with the ecological prudence of farming communities. For this, it is also necessary to spread awareness among the farmers of the ecological best practices. A Village Knowledge Centre (VKC) or Gyan Choupal helps to bridge the gap between scientific knowledge and its field application. It also facilitates the removal of many intermediaries from the marketing chain. India can become a global outsourcing hub in the areas of plant and animal genomics and ICT for rural poor. Farm, veterinary, fisheries and home science graduates have to be trained to become genome and digital entrepreneurs.

The 'Ever-green Revolution' concept is based on a paradigm shift from a commodity centred to a farming system centred approach in technology development and dissemination. The new vision of agricultural growth calls for taking advantage of the opportunities opened up by new technologies and the farmers understanding the importance of integrated natural resources management and ecological gains.

—MS Swaminathan is the Chairman of MS Swaminathan Research Foundation and is considered the father of India's Green Revolution.

Banana is popular for its fibre products. In 2011, the Navsari Agriculture University (NAU) in Gujarat standardised a process of manufacturing high quality paper from banana fibre. Currency notes made out of this paper could last for a century.

(facing page top) Fresh fruit and vegetables on display at a roadside market in Mumbai. In 2010, a leading company launched the 'last mile' solution of refrigeration for the transportation of perishable food items, designed mainly for hot Indian conditions.

(facing page bottom) Indian women work at the Field Fresh Agriculture Centre of Excellence at Ludhiana, Punjab. The centre has an integrated R&D facility with a focus on crop trials and the adoption of appropriate techniques for farming.

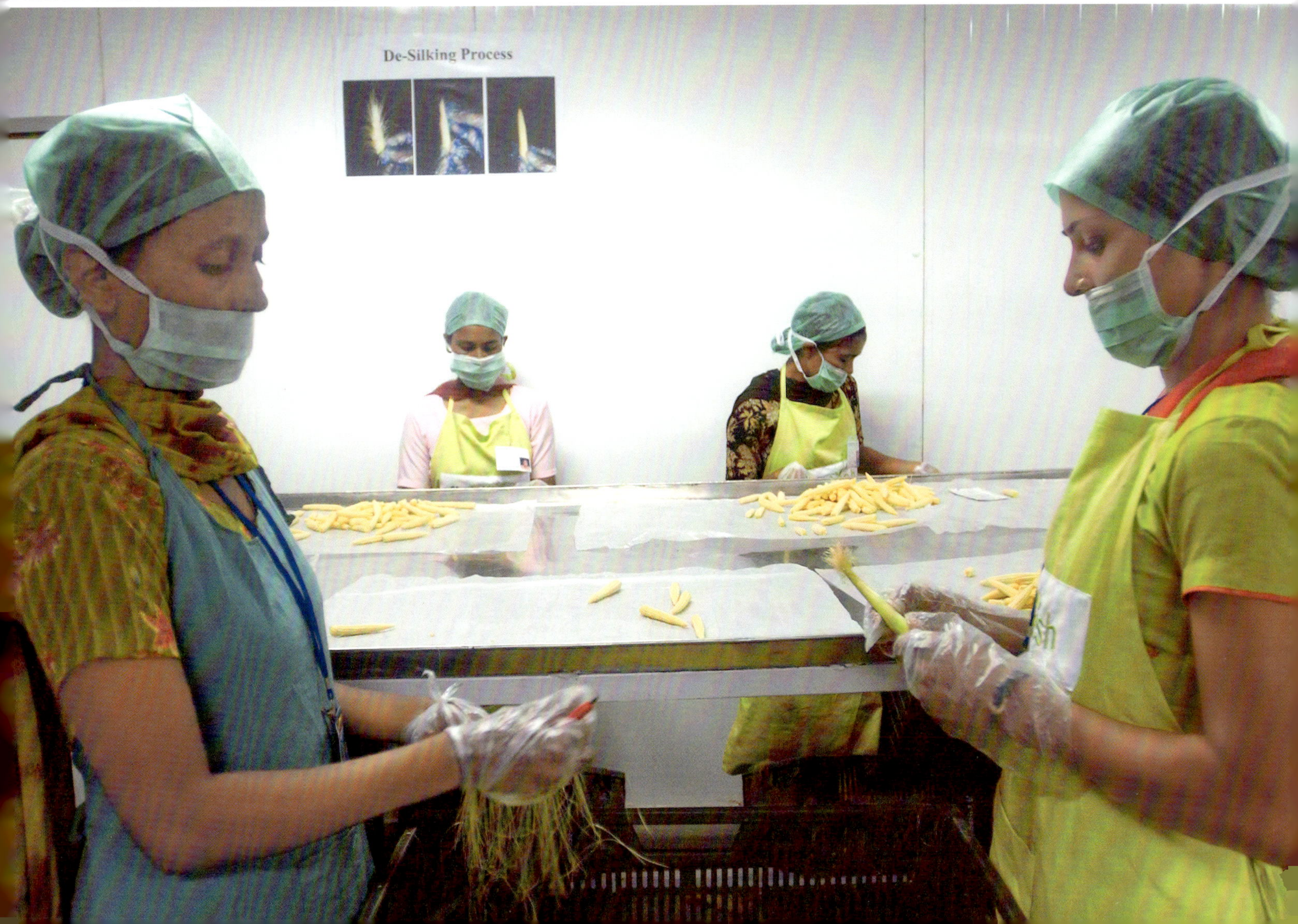

De-Silking Process

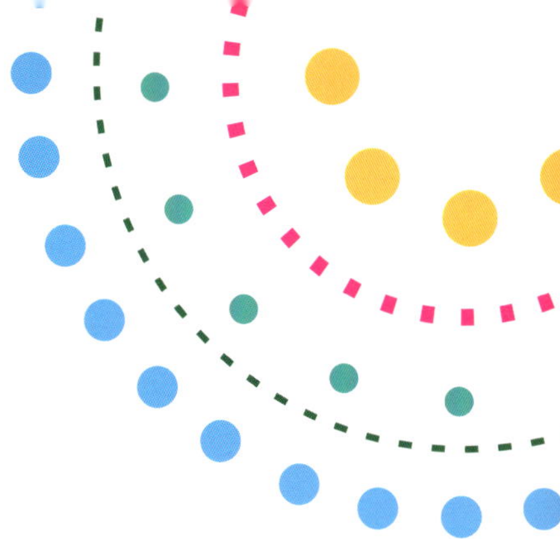

Connect to Decode—
Innovative Crowdsourcing Model

Samir K Brahmachari

Times change and the paradigms that characterise them change as well. Nowhere else is this change more evident than in the world of cyberspace related to the field of scientific discovery and education. The classical model of scientific discovery has almost always needed a laboratory. The conventional model of collaboration and cooperation involves interacting with known person(s). A team is then formed before venturing into the process of discovery. The expertise and the skills of the team members are known in advance and collaborations are pre-defined.

The formal model of education has always operated inside a classroom. There has always been a collection of students and a teacher who instructs and imparts education. The onus is on the student to enroll in the right school, the right college, the right university and to choose the right course. The student must also be fortunate enough to meet an outstanding and extraordinary teacher. However, at present, the entire system of education is perched on the verge of a transformation catalysed by the ICT revolution, which has made access to information, easy.

Earlier accessing knowledge would have necessitated a visit to the library or an interaction with a wise man who would share the information that was sought. Today, we just 'Google' the query! The approach to information-gathering has progressively become more self-reliant. However, concept-formation still remains a challenge.

New Challenges, Newer Solutions

There are a few other, rather more recent, challenges too. For example, how do we harness the power of social networking in bringing together people with expertise in different areas and whom we do not know? How do we harness the power of cyberspace to find inspirational gurus who can motivate the youth to get involved in the process of discovery and to inspire like-minded people to solve a problem that bothers many?

Crowdsourcing is swiftly emerging as an answer to all these challenges. It is the use of open source principles to fields besides software which is an area where open source reported one of its first successes

with Linux. Crowdsourcing is rapidly becoming the foundation stone of a new model for discovery and education. A new paradigm for changing times, crowdsourcing takes the approach to an unheard-of, almost counter-intuitive level by outsourcing collaborative tasks to an undefined group of people (community) who share required expertise and interest. It is a model where a big problem is broken down into smaller, more-manageable modules. Then, skilled and like-minded people are recruited from the cyber world to solve the smaller problems by aggregating their talent and expertise. The underlying principle is enabling participatory efforts of people who may not even be acquainted with each other.

Crowdsourcing is an enormous networking exercise. The combined intellectual power of all the cyber citizens of this connected community defies quantification. Channelised positively, it represents unlimited power to achieve results on an unprecedented scale.

There is little doubt that in crowdsourcing lies the future of tomorrow's science education and research. Connect to Decode is a shining example of an early success of this model.

Connect to Decode (C2D)

The Open Source Drug Discovery (OSDD) is a CSIR led team India consortium with global partnership, with a vision to provide affordable health care to the developing world. Realising the failure of market forces to provide a new and effective drug against Tuberculosis, OSDD began to explore new models of drug discovery; and this was the initiative to which the C2D owes its genesis.

C2D was a project that was launched as a massive initiative aimed to further the understanding of the systems biology of Mycobacterium tuberculosis (Mtb). C2D represented an entirely novel way of tackling a dreaded pathogen, which has evaded mankind's ability to produce an effective new drug that would cure the infection within a short time.

Mtb is the causative organism responsible for Tuberculosis and kills over 1,000 people everyday in India alone and still calls for a nine-month treatment regimen with drugs that were developed fifty years ago. The need was to understand the organism at system-level and to pinpoint its genetic vulnerabilities. It was with this ambitious goal in mind that the C2D programme was initiated. The plan was to re-annotate the biological and genetic information relating to the Mtb genome.

Recruiting Raw Talent

The initial requirement was for a large pool of young, budding scientists to scrutinise available literature, use computational tools and collate this information. Finding willing volunteers was the least of the worries. The day registration to C2D was opened, over 100 volunteers, all undergraduate and postgraduate students, signed in. Subsequently the numbers peaked at about 800. Student volunteers were inducted into the programme and trained over a period of about four months. Their work was monitored online by the respective Principal Investigators. Stringent quality control measures were put in place to verify and correct the annotations made by students who were based in various places.

All volunteers worked under the guidance of faculty drawn from the best institutions. Mentors included scientists from CSIR Institute of Microbial Technology, Chandigarh; Indian Institute of Science, Bengaluru; National Institute of Immunology, New Delhi; Systems Biology Institute, Japan; Anna University-KBC, Chennai; CSIR Institute of Genomics and Integrative Biology, Delhi.

—Samir K Brahmachari is Chief Mentor,
OSDD and Director General,
Council of Scientific and Industrial Research.

The volunteers pooled their time and skills using online tools to provide insights into 4,000 genes of the deadly pathogen. The C2D initiative tackled the Mtb genome from five different angles.

- The first, called TBGO, obtained associations for each gene with function(s).
- Second, each protein (gene product) was studied at high resolution by modelling their three-dimensional structures, through which, higher order clues about their functional roles could be obtained.
- Third, proteins which interact with various sugars or carbohydrates in the cell were deciphered; these are believed to be important for the signalling events in the cell.
- Fourth, the assembly of the proteins was sought to be understood by identifying those proteins that interact with each other directly and those proteins that talk to others through mediators, such as metabolites or other small molecules. It was clear that plenty of experimental studies were buried in literature. The challenge was to manually mine literature to understand each protein expressed by the microbe. The need was to do collaborative literature mining and analysis to unlock relevant information that would be put together to generate the largest metabolic map in System Biology Mark-up Language (SBML) for Mtb genome. The network of the proteins list was reconstructed by integrating data from all the angles. Functional modules (or biochemical pathways) were also identified. The network helps in understanding how cellular activities (metabolism) take place in the bacterial cells. The network provides clues to how the microbe's metabolism compares with humans and with other bacteria. It also helps in answering questions about what strategies should be adopted to kill the bacteria.
- A fifth theme included the identification of proteins that contain antigenic parts in them that could trigger immune responses in the host and thus ultimately help in rational vaccine design.

Results of the C2D Initiative

The project was implemented in an extremely cost and time-effective manner. In just four months the young researchers achieved what would have taken years to do in the traditional manner. In the final round, the best 120 students from across the globe gathered for on-site final annotation to participate in a conference. Each of these students was awarded an e-notebook through philanthropic funding.

The USP of the C2D experience was that this maiden attempt demonstrated what high watermark of excellence can be achieved if the boundless vitality, energy and enthusiasm of the young students are provided just a little direction and guidance. The volunteers of the C2D project would, under the traditional academic system, never have referred to original research papers but relied only on textbooks. However, for the C2D project they voraciously devoured more than 20,000 original research papers, internalising cutting-edge research findings and becoming intellectually enriched in the process.

Under the traditional system of education, undergraduate and postgraduate students do not get any opportunity to learn and exercise leadership skills. However, in the C2D project, the students spontaneously organised into sub-communities and the leadership pyramid emerged smoothly. The students curated the data themselves and demonstrated every trait of good leadership.

Science 2.0, which allows researchers to connect and share information, is about connecting minds with machines. It is not enough to have just the machines. Ignited minds must be in command. We need to brainstorm beyond boundaries and to access the innovation at the fringes. We also need to connect and seamlessly integrate people, information and processes. Only then can we provide end to end integration resulting in a connection of discussion to discovery. C2D did all this and more!

C2D demonstrated the power of people, particularly young people, to connect through the internet, and accomplish complex research tasks. It was also a distinct move from a hierarchical-based model of doing science towards one of equal collaboration.

This was the first time that a comprehensive mapping of the Mtb genome was compiled, verified and made publicly available in an internationally accepted format. This exercise has provided clues towards obtaining a comprehensive view of the microbes as a whole, producing several testable hypotheses that can now be verified in the laboratory.

The C2D project was greatly benefited when Infosys came forward to create a peerless platform that met the most stringent of OSDD's requirements. Infosys designed the Sysborg 2.0 platform to enable scientists around the world, who are dedicated to a common cause, to deliver better quality of work. Today the work is held in a shared database, globally accessible to any research institutions involved in Tuberculosis research. C2D's findings may be used as development opportunities for urgently needed new drugs for Tuberculosis.

Appreciation for OSDD and C2D

The global scientific community has taken note. As Bernard Munos pointed out in a scientific journal, 'Within weeks, 830 volunteered to re-annotate the entire Mtb genome. The work started in December 2009 and was completed by April 2010, packing nearly 300 man-years into four months!' The media response has been encouraging. In its Editorial, *The Hindu* recognised C2D as a 'novel way to do science in the internet era'. Also, the former President of India, Dr APJ Abdul Kalam has lauded OSDD as, '…A valuable open source movement initiated in the Indian health care sector'. It all goes to show that crowdsourcing is a maverick idea for scientific education and discovery that has been noticed. It is here to stay. To stay and to deliver. To deliver and to change the way the world works!

Acknowledgement: *All members of the OSDD community of over 4,500 registered participants from 130 countries.*

Science for Strength, Security, Peace and Prosperity—**DRDO**

VK Saraswat

Needs of a country's war machinery have been the driving force behind a large number of innovations across the world. These have transformed not only the way battles are fought but also made an impact on the lives of the common people. Internet is one such beaming example as it was initially made for the exclusive use of the US defence forces but now the entire world is connected through it.

Formed in 1958, Defence Research & Development Organisation (DRDO), India's premier organisation, with its fifty-two laboratories and dedicated human resource comprising over 28,000 persons, has produced a wide spectrum of technologies and systems that have boosted the country's defence as well as economic strength. Significantly, some of the technologies found applications, either as such or with some modifications, have directly helped the common man.

In the area of affordable health care, products and services such as telemedicine, critical care ventilator, coronary stent, cardiac pacemaker, cytoscan, floor reaction orthosis (FRO), artificial lower limb, use of laser in eye surgery have been developed. FRO works on cantilever principle and is a remedy for a polio patient whose quadriceps muscles are paralysed. When body weight is applied the reaction forces from the floor cause turning movement which locks the frail knee joint. The critical care ventilator, INVENTA has been developed to meet the growing needs in the Indian health care sector. With its user friendly interface, it takes just minutes for anyone to start ventilating a patient and make all relevant and critical parameters available at a glance.

Prototype laparoscopic surgery training simulator is an important tool in the hands of surgeons, providing an effective method of training surgeons in hand-eye coordination. Cytoscan is a personal computer-based system found very useful for mass screening of people for cancer. It uses image processing and pattern recognition technology developed by DRDO for military applications. It is capable of acquiring and analysing the image of a cell with a view to detect and classify cell abnormalities and essentially consists of a PC/AT, microscope, CCD camera, frame grabber and a multi module software specifically meant to detect uterine cervical cancer. The system was used under project TULASI, in the rural districts for screening rural and tribal women for cervical cancer.

(facing page top) Indian Air Force pilots take part in the Republic Day parade in New Delhi. The Republic Day marks the adoption of the Constitution of India, which is when the country became a truly sovereign, republic state.

(facing page bottom) Indian scientist at a biotechnology lab in New Delhi. Biotechnology is being employed for various innovations like bioremediation, which is about using green biotechnology to clean environmental damages of the past.

Apart from making medical facilities affordable, DRDO's innovations have helped the common people of the country in various other fields. Disposal of human waste has been a serious problem in cold Himalayan region as the low temperature hampers the activity of microbes responsible for biodegradation of organic wastes. The partially decomposed mass leads to various water-borne diseases and organic pollution, besides being a nuisance. When the snow melts, the accumulated mass finds its way to rivers and water bodies, disturbing aquatic ecosystem and distributing the pollution to much wider region. DRDO initiated work at its Gwalior-based Defence Research and Development Establishment (DRDE) to solve the problem employing innovative use of anaerobic biodegradation with the help of low temperature adapted inoculum. Two types of bio-digesters, one meant for soil bound low temperature region and other one for glaciers were developed. These bio-digesters degrade organic matter under

anaerobic conditions with biogas (mainly methane) as the main by-product. The anaerobic nature of process leads to destruction of pathogen. These bio-digesters are under operation at the altitudes as high as 18,000 feet.

Besides installation in high altitude civilian areas, in a far-reaching step touching the lives of millions of citizens across the country, Indian Railways has decided to install bio-digesters in trains. A MOU has been signed between DRDO and Indian Railways to introduce this technology. Lakshadweep administration, facing a similar problem due to most of the land having altitude close to sea level has also placed an order of twenty-one bio-toilets. Thus, another technology, originally meant to assist armed forces, has found its way to serve larger interests of the society.

Another important area is the agro-animal technologies developed especially for high altitudes. In the northern frontiers, India has a long border in the high altitude region that includes cold deserts of Ladakh. The region is characterised by very low levels of humidity and soil moisture, low atmospheric pressure and oxygen but abundant sunshine.

An Indian security official poses with an assault rifle at the DRDO pavilion during the fourth Defence Expo in New Delhi. The Expo is attended by more than 400 companies from almost forty countries.

No wonder, in the sixties, the region had negligible vegetation with only four to five types of vegetables being grown in the region. Our forces deployed in this harsh region had no option but to depend mainly on packaged food, particularly during long winter when supply lines remain cut off most of the times. A serious problem, established by scientists, is that lack of proper food adds significantly to the stress levels especially under adverse circumstances. The DRDO took up the challenge and set up Defence Institute of High Altitude Research (DIHAR), to study various problems associated with the region.

The efforts have yielded rich dividends. Agro-techniques for seventy-eight types of vegetables have been standardised to suit the prevailing conditions in a farmer's field in Ladakh. Of the seventy-eight types of vegetables found in the region, farmers in Ladakh have adopted technology for growing twenty-three types of vegetables on commercial scale and are supplying the produce to the army in the region through a cooperative society. Apple and apricot form the major fruit crops in the region. With the support of DRDO, demonstrations of high-yielding, delicious apple varieties have been given to the farmers and every year more than 40,000 grafted seedlings are being distributed amongst them. Several temperate fruit crops have been introduced in the region with success. Farmers have started growing strawberries successfully. Improved cultivars of peach, plum and pear are also being tried as the potential fruit crops in the region. Recent introduction includes blue berry, kiwi fruit and cherry.

Until recently, seabuckthorn, a wild bush was considered a thorny menace by locals, with no commercial value. The plants were being uprooted due to fear of spread to fertile soil. DRDO developed technology for processing this valuable plant with unique nutritional and medicinal properties. The way seabuckthorn was seen changed forever. With the first processing unit established in Leh district of Ladakh region, seabuckthorn collection has become an important activity and an additional source of income in the region. The yellowish-orange fruit is

now popularly known as Leh Berry. It is used for making a number of products ranging from jams, jellies, juices to beauty products and medicinal oils. The collection period is short and the return is high. Seabuckthorn fruit worth ₹1.4 crores was sold in 2007 from Leh, which accounts for less than 5 per cent of the region's total potential. Since collection and primary processing is being done by locals, the revenue generated benefits the needy sections of the society.

Successful demonstration of feasibility of growing these fruits and vegetable crops has been increasingly fulfilling the requirements for local consumption as well as the army. Today more than 60 per cent of armed forces' needs for fruits and vegetables are met by the locally grown ones. The rising prosperity and increasing economic activities in the region are a bonus.

The DRDO does not stop at creating machineries or other gadgets to help the army and civilians. It makes sure that the locals are prepared to face the ever changing world by teaching them about the latest innovative technologies. The DRDO regularly conducts training for unemployed youths on agro-animal and post-harvest technologies. Many of the trained manpower have taken to agro-animal technologies as their sole source of income. Several processing units have been set up in Ladakh by youths who were trained by the DRDO. During the last ten years (2000-2009), farmers in Ladakh have earned an average of ₹25 crores per year with almost 24 per cent annual growth rate by adopting the high altitude agro-animal technologies developed and disseminated by DRDO. In the financial year 2008-09, the farmers of Ladakh earned over ₹41 crores.

Thus, many high end technologies developed for meeting the nation's defence requirement are also directly benefiting the society, especially the weaker sections. Science for strength, security, peace and prosperity is the cherished goal of DRDO.

—VK Saraswat is the Scientific Adviser to Defence Minister of India, Secretary to Govt of India for Defence R&D and Director General DRDO.

A marathon runner during the 'Great Tibetan Marathon' in Leh, the capital of Ladakh. Nearly 200 marathon runners from seven European countries, Australia and India completed the 42-kilometre run—the world's highest altitude marathon. New agricultural techniques and practices have been employed in the region, one of which has been to cultivate the Leh Berry. The wild fruit is used to produce jams, juices and jellies, among other products.

PSLV
C9

Social Innovations of
Space Programme

K Kasturirangan

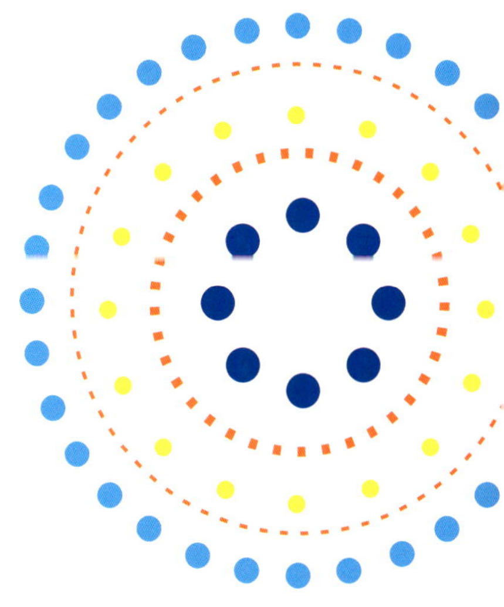

Our tryst with space has taught us, in a humble way, that the enterprise is essentially a people's enterprise and it must always orient its activities towards the mainstream of the developmental march. It demands the best of our qualities and a quest for excellence and excellence alone.

I nnovative societal applications of the space technology spring from its unique characteristics—its ability to cover large areas of earth's surface from a vantage point and its capacity to bridge the divides, irrespective of remoteness of the locations and distances. It can be used in an area, irrespective of the prior state of development there, despite the absence of terrestrial infrastructure. The space technology has high resilience to natural disasters. Its capacity to converge with a range of complementary technologies makes it an invaluable tool to assist social and economic transformation.

Dr Vikram Sarabhai, the architect of India's space programme, provided the vision for self-reliance and the use of advanced space technologies for accelerating national development. Through multifaceted innovations in technology, organisational systems and applications, over the past four decades, ISRO has acquired capabilities to build and operate many state-of-the-art space systems and apply them to development in sectors as varied as telecommunications, broadcasting, weather observations and natural resources management. India now claims a vast array of applications of space technology for development.

Institutional Innovations
Experience from earlier experiments involving broadcasting, communication and remote sensing, and dealings with the user communities provided inputs for the creation of appropriate formal institutions. For example, remote sensing experiments led to the establishment of the Planning Committee of the National Natural Resources Management System. Such a structure enables the involvement of major user communities and incorporation of new and powerful techniques into the related conventional system. Similarly the multi-agency INSAT Coordination Committee, as the apex body, addresses the development and use of space communication and broadcasting technologies. The broadcasting capability has been used through the same mechanism to promote distant education, developmental communication and other socially relevant

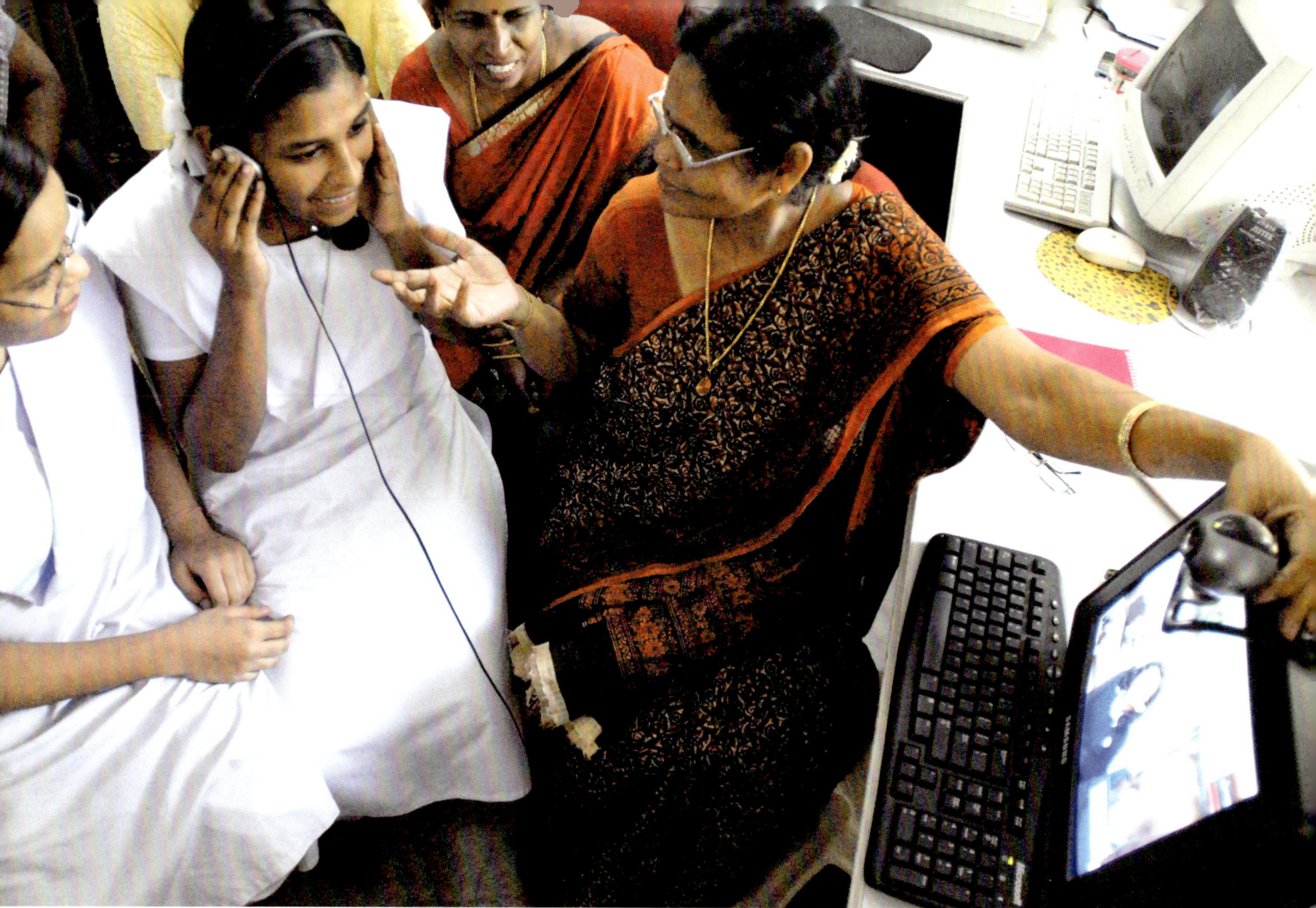

developmental activities. Such institutional structures are unique to India. As the expanding space activities create new opportunities for the NGOs, industries and service organisations, ISRO promotes the use of space infrastructure through its commercial arm, Antrix Corporation. The association with non-governmental agencies in areas such as watershed development and telemedicine has already yielded significant results.

Space Application Innovations

The Indian Remote Sensing (IRS) satellite systems have made considerable impact in the areas of natural resources management, environment monitoring, disaster management and infrastructure development. Several Earth Observation (EO)-based national missions such as Rajiv Gandhi National Drinking Water Mission, wasteland mapping, watershed monitoring projects, crop acreage and production estimation, satellite-based potential fishery zone assessment have provided

inputs for the policy, developmental planning, monitoring and evaluation.

The EO satellites, combined with tools like Geographic Information System (GIS), provide timely and cost-effective inputs for decision-making in natural resources management. Such projects generate community centric EO products such as maps for groundwater prospects and recharge sites, delineation of wastelands for reclamation programmes, action plans for rural employment and income generation schemes.

Participatory Watershed Development

Watershed development targets soil and water conservation, afforestation, horticulture and other vegetative treatment. Space applications have been tailored to respond to such requirements and facilitate participatory approach for watershed development.

Sujala, is a community driven, participatory, integrated watershed development programme, implemented in five rain-fed districts of Karnataka: Kolar, Tumkur, Chitradurga, Dharwad and Haveri covering about 0.5 million hectares encompassing about 1,270 villages benefitting about 3,50,000 households. It is an example of effective use of space

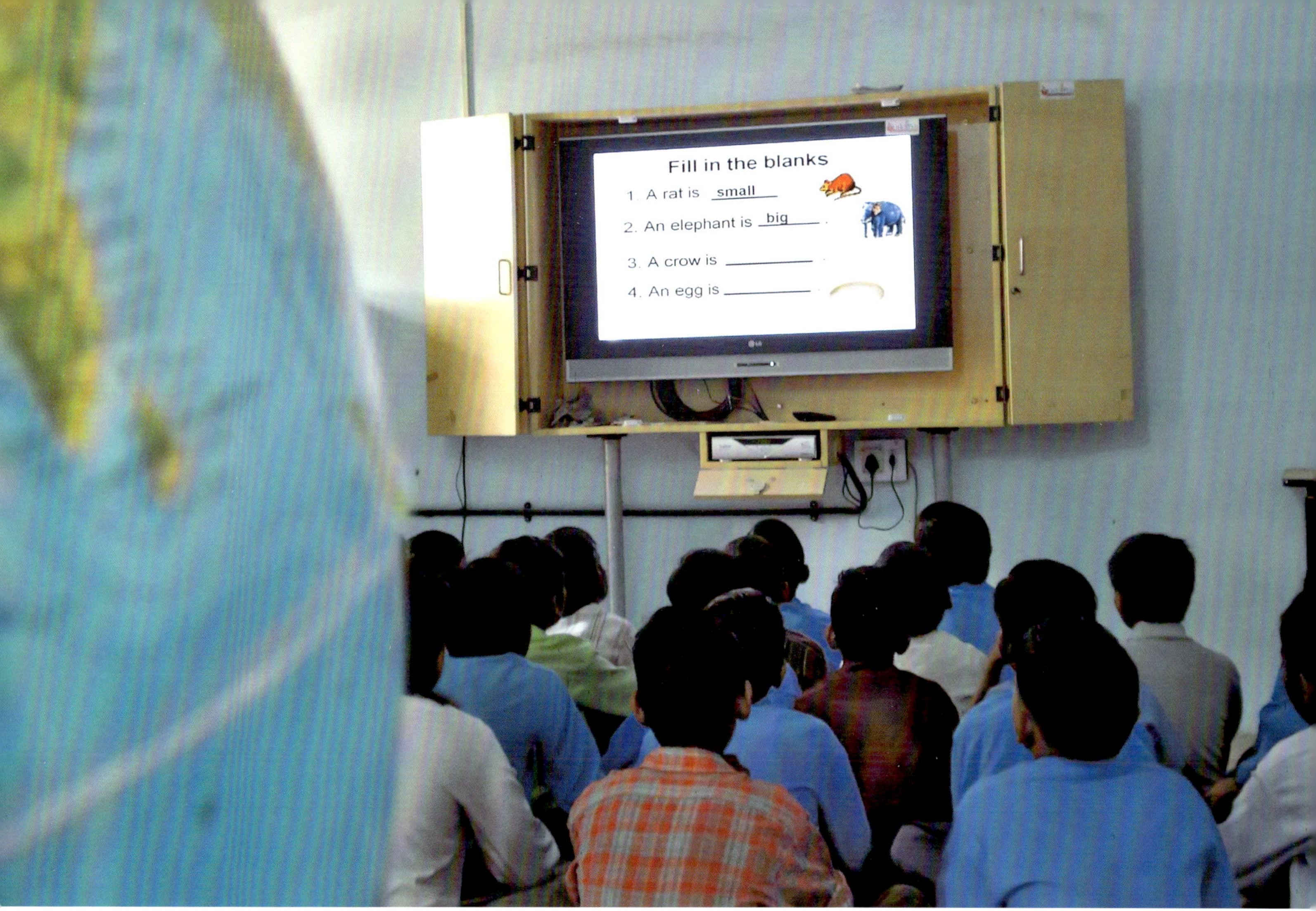

technology for watershed development. The methodology involves the use of suitable space imageries, creating layers of information from the same such as soil, groundwater, surface water, geology, vegetative cover, etc., integrate the same with toposheets and ground truth information, and thus create a GIS data base. When modelled, such data base could provide optimal solution for sustainable use of land and water in terms of productivity. Sujala was implemented to alleviate poverty in predominantly rain-fed areas by improving the productive potential of selected degraded watersheds, enhancing production and livelihood systems and strengthening community and institutional arrangements for better natural resource management. It has helped rejuvenate the natural resource base, reducing the runoff, arresting the soil erosion and improving the productivity. The project has not only increased community participation and awareness, but has also led to women empowerment, better standards of living and health and reduced migration. The impact assessment showed an improved cropping intensity, cropping diversity, crop yield, and an increase in groundwater levels.

Crop Monitoring

The advancements in present satellite sensors help in arriving at the pre-harvest acreage estimates of major crops. Remote sensing is helpful in yield modelling. Generally, statistical, meteorological or spectral models are used for crop yield estimations. The space technology has made a revolutionary impact on spectral models. The availability of cloud-free optical sensor data was a major challenge especially during the kharif season. ISRO has developed techniques using microwave remote sensing data to overcome this limitation. The space technology innovations enabled yield forecasts which are available one month before harvest.

Land Degradation Mapping

The multispectral data gives information on the extent, location and magnitude of less productive lands such as eroded lands, salt-affected soils, waterlogged areas and lands affected by shifting cultivation. This information has been used for

A boost in distance education. In 2004, ISRO launched EDUSAT to create interactive classrooms in the remote and rural areas of the country, ranging from primary education to vocational courses. The aim is to establish the first e-university in India.

GHA-TROPIQ
SPACE RESEARCH ORGANISATION
MEGHA-TROPIQUES

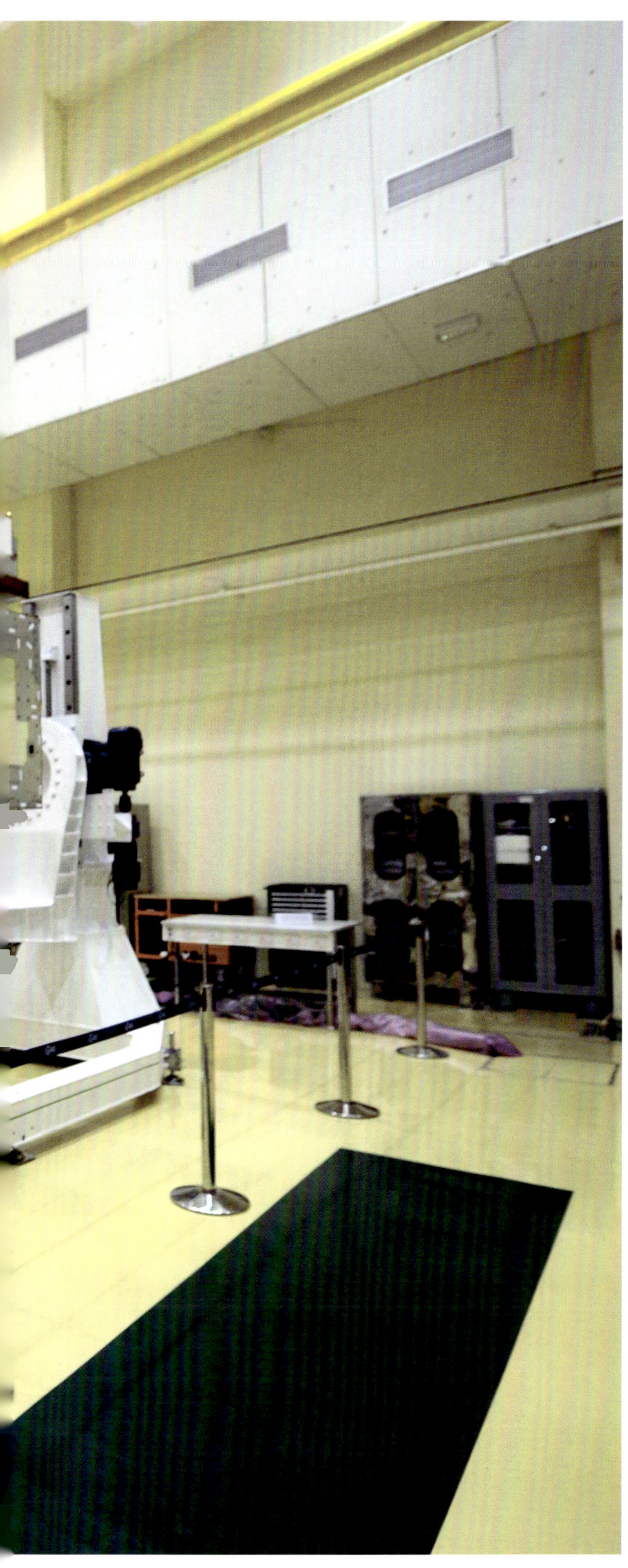

planning land reclamation and soil conservation programmes.

Forests and Ecosystems

Remote sensing through satellites has enabled generation of forest cover maps. The management of forest resources in India is based on a three-tier monitoring system. At the national level, biennially forest cover mapping is carried out using satellite data by the Forest Survey of India on 1:250,000 scale. Relevant inputs required for forest management at the state level are generated at 1:50,000 scale or even at 1:15,000 scale. The data from the high resolution Indian satellites is supplemented with that collected through aerial photography. The grassland mapping is another area where remote sensing has played a key role.

Water Resources Management

The spectral measurements taken by satellites are extremely useful for diverse needs of water resource management covering rainfall estimation, snow and glacier studies including snow melt runoff forecasting, irrigation command area monitoring, reservoir sedimentation, water quality assessment, and groundwater assessment and prospecting. Several case studies of command area development, groundwater inventory, canal alignment, irrigation performance evaluation and flood mapping, drought assessment, have proved that an integration of remote sensing and conventional approach significantly decrease the cost and time involved in projects.

Under the National Drinking Water Technology Mission, the Department of Space, Government of India with the active cooperation of various user departments has prepared district-wise hydro-geomorphologic maps on 1:250,000 scale covering 447 districts in the country using satellite imagery with limited field checks and available information.

Ocean Applications

The use of high spectral resolution ocean colour monitor helps identify the phytoplankton distribution, which in turn makes it easy to locate potential schools of fisheries. The sea surface temperature and Potential Fishing Zone (PFZ) maps

An Indian official stands beside the Franco-Indian Megha-Tropiques satellite in a 'white room' at the ISRO satellite centre in Bengaluru. The Megha-Tropiques Mission is a planned mission to study the water cycle in tropical atmosphere, in the context of climate change.

have been generated routinely using data from OCEANSAT and other satellites. Currently, the PFZ maps are generated biweekly and disseminated to fishermen across the coastal areas by the Ministry of Earth Sciences. The PFZ maps have helped fishermen to get better fish catch. The OCEANSAT data is used for generating information on wave energy and directional spectrum, which are essential for coastal zone development and execution of offshore projects. The Indian coast has been mapped using IRS data, enabling the protection of coastline from degradation and to ensure optimal land use. Coral reef mapping has helped in the protection of active coral reefs.

Ocean circulations are the engines of weather phenomena. Sea topography variations derived from altimeter, sea surface winds from scatterometer, sea surface temperature from satellite mounted radiometers and salinity from microwave radiometer are some of the inputs for ocean circulation prediction models.

Disaster Management

In the event of natural calamities like drought, flood and cyclones, that adversely affect the water security, space technology has made substantial contributions in different phases such as preparedness, prevention and relief. While geo stationary satellites provide continuous and synoptic observations over large areas on weather including cyclone tracking, the polar orbiting satellites have the advantage of providing much higher spatial resolution in images that could be used for detailed monitoring, damage assessment and relief management. The Disaster Management Support programme uses space communications and remote sensing capabilities. The relevant ISRO centres are networked with the National Emergency Control Room and State Control Rooms through satellite-based, secure Virtual Private Network, which during the period of natural disasters help in establishing real time network solutions for data transfers at various levels, empowering the vulnerable communities.

Tele-education

ISRO's initiatives for using satellites for education in both formal and non-formal domains are noteworthy. Interactive satellite terminals have been set up in the remote classrooms of various states. Under tele-education programme, more than 50,000 classrooms cover primary education, secondary education, professional education, adult education, teachers training and non-formal education.

Telemedicine

ISRO has also shown how health sector can benefit using modern satellite technologies—by using satellite communications to bridge the 'gap' between a patient in remote areas and a doctor or specialist in urban areas. Initially, telemedicine system has been configured by ISRO with 'point-to-point' connectivity between a village hospital and the speciality hospital situated in a city. Subsequently, server and browser-based telemedicine system has

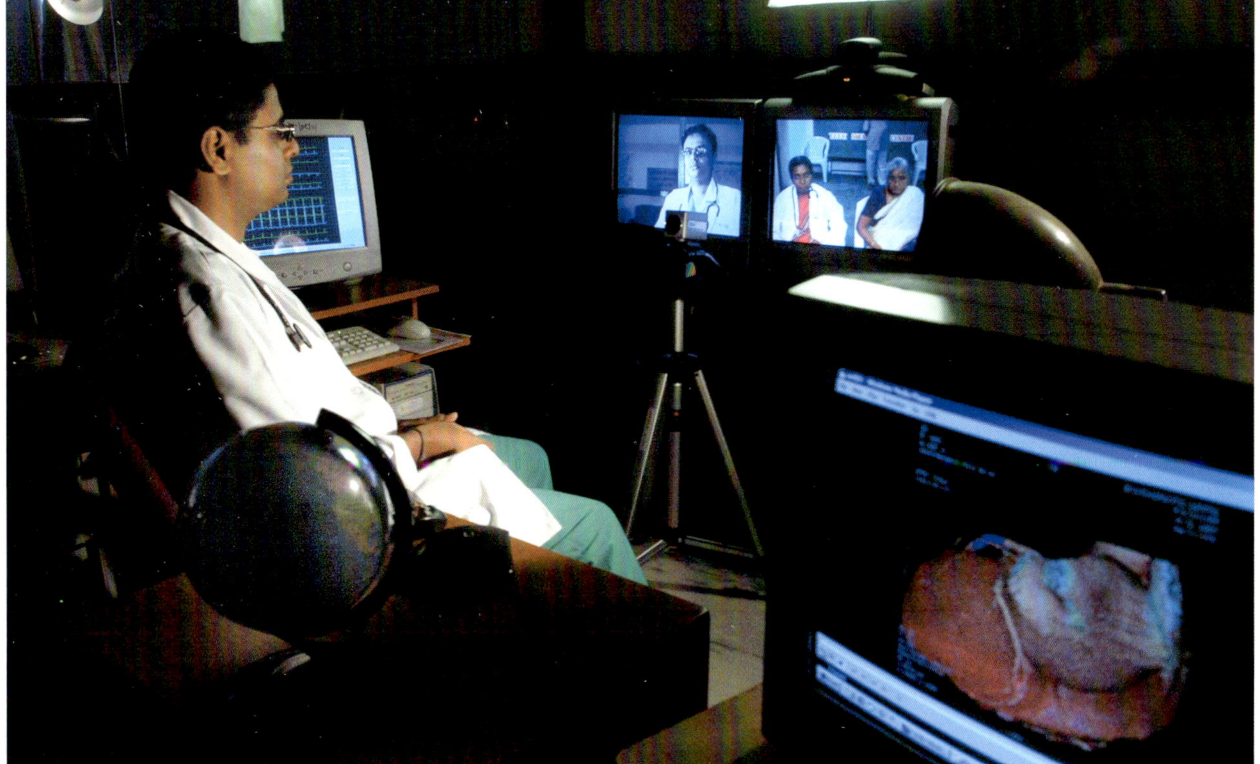

been evolved for multipoint connectivity and the same is adopted for connecting a remote rural hospital with any super speciality hospital among a large pool of them.

Some states have already introduced telemedicine in an operational mode and have prepared the district hospitals with telemedicine facility both for ambulatory and intensive care for cardiac related treatment. Satellite-based telemedicine facility has been made operational so far in 382 centres. Mobile health vans with real time connectivity have taken telemedicine to some of the remotest places.

Village Resource Centres

In the present context of development in our country, it is imperative that no one should remain devoid of knowledge. ISRO has developed the system of Village Resource Centres (VRCs) to empower the masses. VRCs provides health services, skill centric non-formal education, geo-spatial information on natural resources and disaster vulnerability, and weather and market-based advisories to farmers for whom such information is critical.

Tele-education (mainly non-formal) is provided through appropriate content development and live interactive programmes. The satellite-based telemedicine programme has given a new dimension to the rural health care in India. Thanks to the telemedicine networks, the local doctor from Primary Health Centre can go to the Very Small Aperture Terminal (VSAT)-based hub facility for expert consultations with a specialist in a city hospital.

These VSAT-based hubs also provide local weather related information. The interactive VRCs have been set up in 479 locations in association with NGOs. These have taken the benefits of the space technology to the doorsteps of rural population by providing telemedicine, tele-education, disaster management support, advisories to farmers and fishermen, through a single window system.

The Indian space programme has enabled us to learn several things, quite unique in their culture, content, approach and of great grass-roots relevance. The success of space applications signals the society's capability to bring about a synergy between highly diverse systems of high-technology, applications and development, a capability characteristic of economically and socially advanced societies.

*—K Kasturirangan is the Member,
Planning Commission, Government of India.*

Telemedicine is being put to significant use in India, particularly for people living in rural areas. With the 3G communication system being introduced in the country, the telemedicine sector is all set to receive a boost.

(pages 60-61) In 2009, ISRO launched OCEANSAT-2, country's sixteenth remote sensing satellite to identify potential areas for fisheries and to forecast critical inputs on weather and climate change.

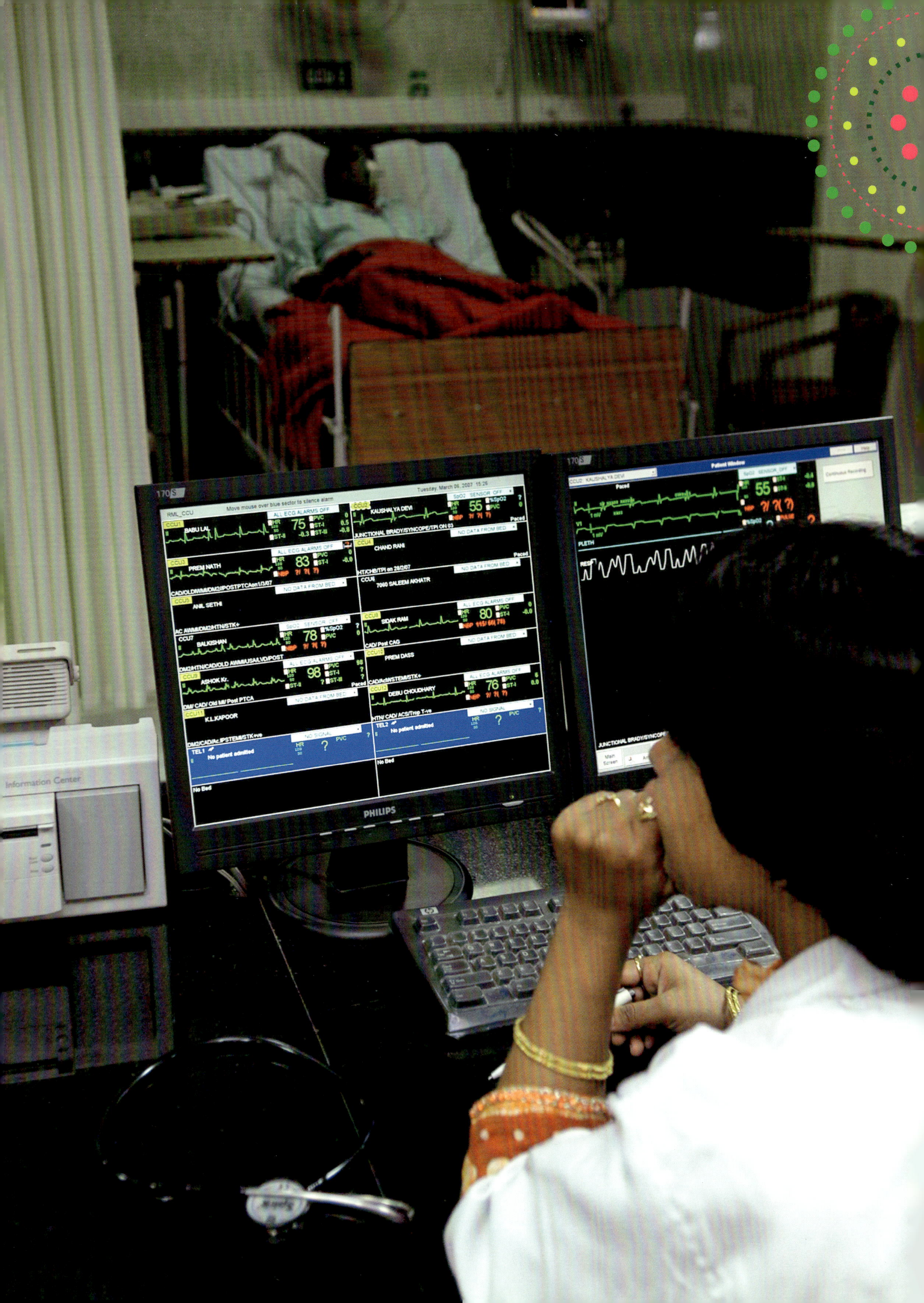

Societal Benefits from Developments in **Atomic Energy**

Anil Kakodkar

Mention the word 'nuclear' and the images of the bomb or a power station featuring a reactor flash in popular imagination. Lacking in visibility are the various uses of atomic energy in the fields of food and agriculture, water, health, industry and waste management. Such applications mark a distinguishing aspect of India's atomic energy programme. Those who envisioned India's plan provided for harnessing nuclear energy for societal benefits even in areas other than power, with the resultant impact in several areas.

India's atomic energy establishment pursues basic and applied research and activities related to technology development, demonstration and deployment in an integrated manner. It encourages practically all areas of scientific research primarily based on the interest of scientists. Putting together a competent interdisciplinary team to address a particular technology development challenge is thus easily done in the Bhabha Atomic Research Centre (BARC). The spin-off technologies bring unexpected benefits in various sectors. The work done as a part of mission objective also gets applied in other areas with marginal additional effort towards customisation. In order to realise the full potential, a mechanism to involve resources of partners interested in carrying the technology forward for commercial use has recently been set up. This should considerably enhance the benefits to industry and society from the atomic technology in the years to come.

Development of improved crop varieties through an effective blend of mutation and recombination breeding has been a huge success story at BARC. A relatively small group through a comprehensive collaborative programme with a number of agricultural universities and other agencies has developed as many as thirty-nine new varieties that have been released and Gazette notified by Government of India for commercial cultivation in different parts of the country. Oil seeds and pulses, which continue to be in short supply, form a major (thirty-seven out of thirty-nine) share of this development. Estimated share of BARC developed varieties in the national production could well be around 20 to 30 per cent for groundnut, around 20 to 30 per cent for black gram and around 10 to 15 per cent for moong bean. On an average three new

varieties have been released annually over the last five years. This trend is expected to continue; extending to other crop varieties as well.

Radiation is fast becoming an important tool for enhancement of shelf life and hygienisation of food and medical products. Importance of this technology was realised in the aftermath of food poisoning outbreak caused by E.Coli 0157:H7 and other epidemics caused by food borne pathogens. In India, this technology has an important role in widening the market access for perishable food and other agro-products, reducing wastage due to spoilage, ensuring ready availability of sterilised medical products even in villages and remote areas and enhancement of export of hygienised products that meet quarantine and health standards of importing country. Number of such radiation processing facilities have registered a steady growth (around two per year) in recent years and has reached fifteen as of now. Together these plants may well have a processing capacity of around 10,000 tonnes per year. Recent opening up of mango exports to USA was possible as a result of this technology.

Apart from processing food and agro-products as well as medical products, radiation has been found useful in several other important applications for processing value-added natural and other products. Radiation processed hydrogel for burn and wound dressing has been one such development. It can hold large amount of water within its structure, hence when applied on a wound it has a soothing effect which relieves pain of the patient.

Another remarkable thing about this product is that unlike other dressings, it does not cover the wound. Hydrogel dressings are transparent and hence the healing of the wound can be monitored from time to time, without giving any discomfort to the patient.

Even in the area of medicine, radiation and radio-isotopes are playing an important role. Around 180 nuclear medicine centres in the country perform more than 0.3 million in-vivo diagnostic and therapeutic procedures annually. Apart from that, there are around 200 radioimmunoassay laboratories that carry out more than 2.5 million assays annually. Most of these centres are being supported with supply of radio-pharmaceuticals processed by Board of Radiation and Isotope Technology (BRIT) using radio-isotopes produced in reactors of BARC. Bhabhatron teletherapy system for treatment of cancer has been an important success story. The system which has been developed through joint efforts of BARC, Tata Memorial Centre (a Department of Atomic Energy aided institution) and M/s Panacea, has turned out to be a state-of-the-art system that in some respects is even superior to the best machine anywhere. The country needs several hundred such machines to meet the requirements of cancer treatment in district hospitals and other locations in the interior where one needs robust products that can work under conditions of poor quality of power supply and inadequate hi-tech maintenance support. Already twenty Bhabhatrons have been installed in different

A young woman at work in a lab. The nuclear medicine centres in India use radio-pharmaceuticals manufactured by Navi Mumbai-based BRIT to diagnose and treat cancers and thyroid disorders.

hospitals in the country including one in Vietnam. Besides making available a world class teletherapy system at much lower prices, this development has in fact led to significant lowering of market prices of similar such imported systems.

Sewage sludge which contains most of the pathogens after treatment of sewage is a major burden on the urban health management system. By hygienisation of this sludge using radiation it can be turned into a major source of safe manure for improvement of soil quality and resultant agricultural productivity. The Sludge Hygienisation Research Irradiator (SHRI) set up at Gajrawadi sewage treatment plant of Vadodra Municipal Corporation has been a great success story. The farmers in the neighbourhood have organised themselves to derive maximum benefit from this technology.

While the above developments are based on Cobalt as the radiation source, electron accelerators have also become viable and precision sources of radiation and are increasingly being used for various radiation applications. In order to promote the technology for manufacture of such machines and their applications, an Electron Beam Centre has been set up jointly by BARC and Society for Applied Microwave Electronics Engineering and Research (SAMEER) at Navi Mumbai.

We are fast heading towards a serious water crisis. Addressing issues related to availability of water as well as its contamination need appropriate technologies that are affordable. In this context BARC has set up a 6.3 million litres per day capacity hybrid desalination plant at Kalpakkam, Tamil Nadu based on reverse osmosis and multistage flash distillation. This unique facility is coupled to Madras Atomic Power Plant and provides mineral free water for process use and blended water with controlled mineral content for human consumption. BARC is also involved in guiding a number of different agencies in setting up desalination plants as well as in water shed management using isotope hydrology. At a completely different level water purification techniques developed at the centre have found widespread applications in urban and rural areas for safe drinking water.

NISARGARUNA, a robust biphasic bio-digester developed in BARC provides an effective decentralised solution to management of bio-degradable urban and rural waste including excess agricultural residue, in an environmentally friendly manner. This zero waste technology eliminates the need for long distance transportation and land fill, reduces burden on health management infrastructure, produces energy and soil enriching manure and provides livelihood for the poor. Around 120 such plants are already operational in different parts of the country with sixty more plants under construction and another 100 that have been sanctioned. Operational plants have the capacity to process 450 tonnes of waste everyday.

Having developed several technologies of use in rural areas, BARC is experimenting with a new model for technology oriented rural development. It has helped set up Advanced Knowledge and Rural Technology Implementation (AKRUTI) centres at twelve locations in the country. These are demonstration centres for local innovation that customise the laboratory development for local fabrication and use. These self-sustaining centres create a pull for such ideas in the neighbourhood leading to self-empowered development. While development of appropriate technologies requires convergence between developers and users, there are other players who also need to provide a constructive support. They include NGOs that work at the grass-roots, government agencies that provide all important policy support and financial institutions that enable credible deployment.

Thus, in the context of atomic energy it is imperative to mention that India is not developing technology to become a world leader, but is doing so to meet the aspirations of its people.

—Anil Kakodkar is the Former Chairman of the Atomic Energy Commission, Government of India.

FOCUS
IX20
IX50

रचनी मेंहदी
28
25
34
दलिया स्पेशल
Rs. 16
मसर काली
मोठ स्पेशल
चना दाल
35
उड़द सबुत
काबली डालर
44
मक्का लाल
30
राजमा लाल
32
अरहर कानपुरी
मटरा हरा
Rs. 22
मटरा सफेद
Rs. 20
अरहर बडा

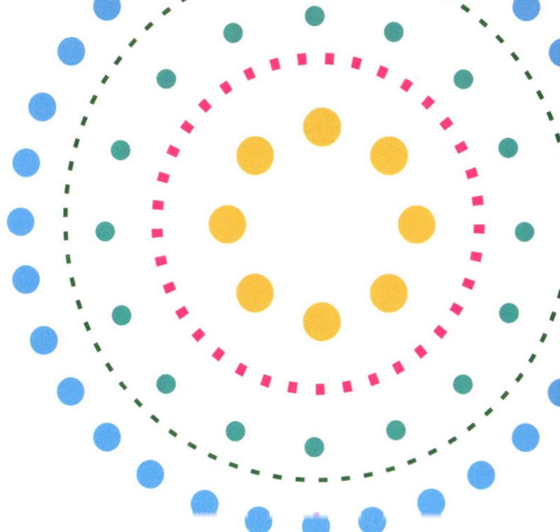

Innovativeness:
the Tata Experience

R Gopalakrishnan

American evolutionary biologist Stephen Jay Gould pointed out that in nature, change occurs discretely, not continuously. He called it 'punctuated equilibrium'. In the Tata group too, it appears that innovations have demonstrated such punctuated equilibrium. Some innovations in the group confirm the influence of an individual, for example, a visionary leader. However, there also seems to be at play a democratised form of innovation through which groups of operating level managers have innovated.

In such an environment, many employees feel a sense of involvement and are motivated enough to contribute their best. This leads to continuous innovations achieved in an apparently effortless way. Such innovations do not get noticed widely, unlike the leader-driven innovations that bear the mark of individual brilliance and are written about and eulogised.

Academics have described four techniques in regard to opening up new and large-volume market spaces—create exceptional utility, set a strategic price, build a profitable business model and overcome adoption hurdles. Ginger Hotels (Indian Hotels) and the Ace (Tata Motors) fall in this category.

Process innovators achieve a defined end through a completely different method; for example, a new way to make sheet glass or titanium, the manufacture of bio-fuels out of algae that absorbs carbon dioxide, and so on. Tata Chemicals is an example of a process innovator—the company uses a novel separation and filtration technique for waste effluent that has led to the production of a new end product—cement.

Business model and management system innovations are about the implementation of new practices, processes and structures that represent a significant departure from current norms. Economist Joseph Schumpeter argued that the process of management innovation is probably as important to economic and social progress as technological innovation. Management system innovations are more difficult to justify and evaluate than product innovations. That is why management system innovators sometimes say that they innovated in spite of the system. Tata Consultancy Services' (TCS) creation of the 'offshore delivery model' in the late sixties is a management innovation; so is the innovative funding model that enabled companies like Tata Tea and Tata Steel to acquire international companies that were three times their own size.

(facing page) At the Geneva Motor Show held in Switzerland in 2011, the Tata group displayed another version of the Tata Nano for the European market, called Tata Pixel. With an innovative diamond-shaped door system, the car is the most efficient four-seater in the world.

The Tata group's organisational structure itself is unique in that it is the converse of traditional structures around the world. It has an unlisted parent with listed operating companies. This enables the group to combine a long-term business perspective without compromising the short-term view across multiple industry segments.

Social application innovation is, as the name suggests, innovation in the social arena rather than the classical business arena. For instance, the innovative model followed by the Tata Relief Committee (TRC) to reach succour to disaster-hit areas. TRC is not a standing committee; it is a self-forming group that comes together whenever the need arises.

For an innovation to be considered disruptive, it should be either novel or should have an impact. Novelty is defined by a context or environment. For example, something may be novel in a country, region or in a particular social context. Novel innovations may be based on a unique technology and can usually be judged in a short time frame—such as Titan Industries' innovation in developing the world's thinnest watch, the Titan Edge.

The impact is defined by the footprint of the innovation, such as a modification in the trading system, a change in customer habits or an impact on the surrounding community. Impactful innovations are better judged with the passage of time. For instance, the launch by Tata Tea of pouches of garden fresh tea in a market inundated with rectangular packets of factory-blended tea is an example of how a company fundamentally changed its business model and the consumer perception of a very old and traditional product.

Adaptation and entrepreneurship are essential manifestations of innovation. Older, well-established companies have been associated with four attributes:

- Adaptability: Great sensitivity to their environment.
- Oneness with society: Cohesive, with a strong sense of identity.
- Decentralisation: Tolerant of activities in the outliers at the edges.
- Prudence: Conservative in financing.

The Tata group definitely displays some of these characteristics, arguably all. It has interesting long-

living entities. Indian Hotels (the Taj brand) is over 105 years of age and may well be the oldest company still owned by the original promoters that is currently listed on the Bombay Stock Exchange. So too is the case with Tata Steel (more than 100 years old) and Tata Power (97 years). The promoter company is Tata Sons. It is unlisted, over 130 years of age and is uniquely owned by charitable trusts to the extent of two thirds of its holding.

There are two very different ways of viewing organisational innovation. The most common is the programmed way where you have to see something driving it and the organisation responding to the drive. You change practices within an existing organisation through learning and adoption of new techniques. The large volume of literature on innovation caters to such companies. This approach may lead to incremental innovations quite often, though not necessarily so. The other way of organisational innovation is the natural way, where you cannot easily see the driving force and the response. There are organisations that just grew up that way; they do not seem to programmatically try to innovate, they just innovate. They seem to have innovativeness as part of their gene code and are not too conscious about it.

Is Tata innovative in any unique way? A unique feature of Tata is the breadth of innovations that have over a long period of time straddled the top and bottom of the pyramid: for example, creating a super-computer while simultaneously using throwaway laptops to promote adult literacy.

In the Tata group, there is no evidence of a central directive or a centrally driven methodology for innovation. Each case suggests its own lessons. Possibly a unique organisational paradigm has evolved in the group wherein the tremendous plurality of Tata is embedded within an overarching framework of the organisational aspiration—that the community should be the ultimate beneficiary of its commerce. Perhaps Tata has innovated in a natural way and the aligned aspiration of community benefit has had an over-arching influence.

Managers and organisations use certain metrics to evaluate corporate innovation, such as the number of radically different products launched, research expenditure, number of patents, innovation awards won, and so on. All these measure innovation of the

two types—product and service and process innovations. It is true that in the Indian context, Tata Steel was the first to set up an R&D laboratory (1938), Tata Motors has the highest spend on R&D amongst Indian companies, and TCS set up an R&D laboratory in Pune as early as the eighties when software was not even recognised as an industry. These organisations have produced terrific examples of product and process innovation.

Still, if one relies entirely on such metrics, the strength of the Tata group as an innovator is not very evident. There are hard-to-replicate business model and management system innovations and social application innovations. In these two areas, the Tata group's impressive record of innovation is matched by few corporations in the world. And now that the context of the Tata group's evolution in an inward-looking economy has changed, its companies are responding appropriately. They are acquiring patents and following other global innovation practices. In the coming years, this will enable the Tata group to make a global impact through science and technology.

—R Gopalakrishnan is Executive Director, Tata Sons.

(page 70) A foundry worker at the Tata Steel Limited's Jamshedpur plant. In 2007, the company acquired the Anglo-Dutch steel firm, Corus. Tata Steel is a major global steel player, with a workforce spread across four continents in the world.

(page 71) Tata Nano was the biggest crowd-puller at the Auto Expo of 2008 in New Delhi. Priced at ₹1,00,000, the Tata group gave the world a model of 'disruptive innovation' and the car became an ingenious marvel of the Indian automobile industry.

(top) Tea produced in India is popular across the globe. In 2009, ISRO and the Tea Board of India launched a novel project to map tea-growing areas, with the aim of bringing more area under cultivation in West Bengal and Assam.

(facing page) Mahatma Gandhi adopted innovative, non-violent means to resist the social ills prevalent in Indian society and pave the way for India's independence. Such was the strength of his ideas, that they remain relevant even today and are emulated worldwide. Here he is seen breaking the salt law at Dandi in 1930, which was another expression of resisting the policies of the British Raj. In 1983, the Tata group launched packaged iodised salt in India for the first time.

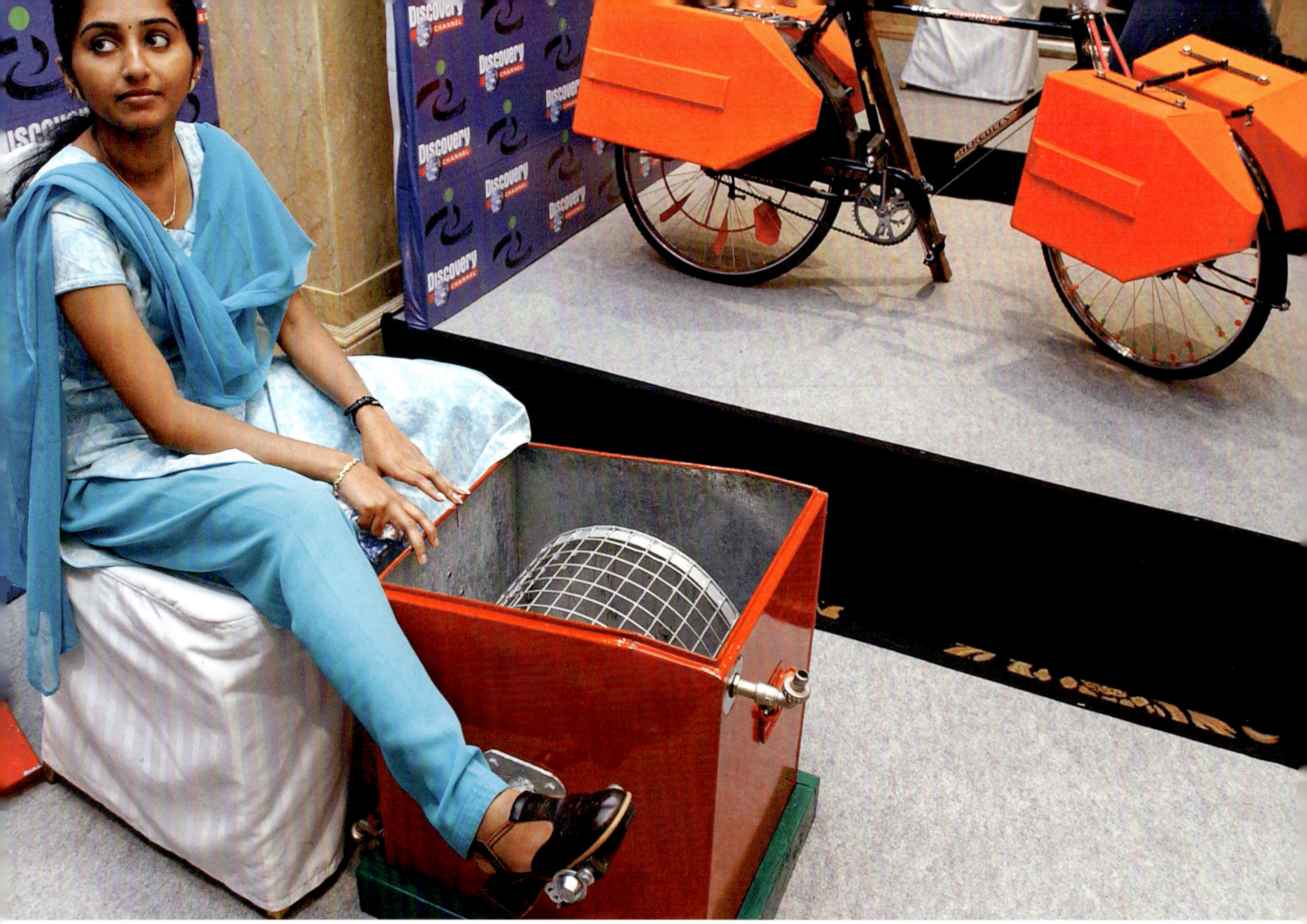

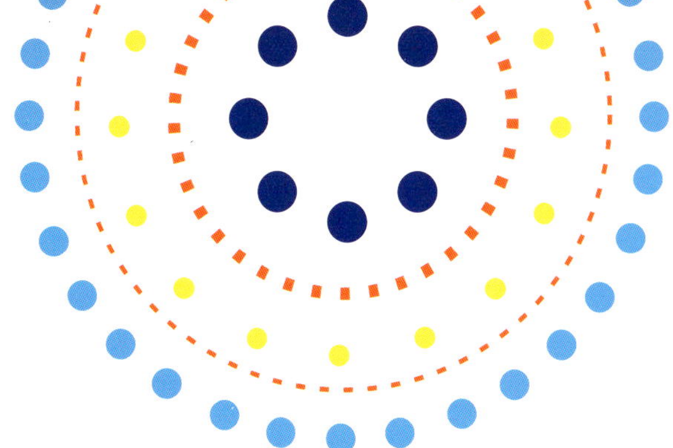

Towards Creative, Compassionate and Collaborative India:
Honey Bee Network in Action

Anil K Gupta

Indian society is going through a transformation as the traditional inertia is giving way and innovations are slowly being recognised as instruments of empowerment and change. The Honey Bee Network has been engaged in an unusual endeavour of collecting through its volunteers grass-roots innovations and traditional knowledge practices. The task is undertaken during two *shodh yatra*s a year. These journeys to the homes, workshops and farms of grass-roots innovators have yielded significant results.

India's heritage is rich in the examples of innovations and expertise in various fields like metallurgy, architecture, water management, health, and food which have stood the test of the time. But over a period of time, we internalised the constraints and generated a culture, which reinforced compromise, compliance and conformity rather than dissent, diversity and innovation. This habit could survive because of an accompanying culture of '*chalta hai*', which means 'everything is all right'. However, the present young generation is refusing to put up with such an attitude. The Honey Bee Network has played a small role in bringing about change in the mindset.

Let me first share how it all started. In 1985, I went to Bangladesh after a bureaucrat read a paper of mine on the sociology of land-use planning and spent an year helping scientists learn from poor people. I discovered such a great creativity among the tenants and landless farmers of Bangladesh that I was completely overawed. I was handsomely paid for the project, that's when I asked myself a question: Did I get paid so well because I am a bright, brilliant professor, or because the people whom I wrote about were brilliant? If it was the latter, then logically my income-tax returns should have reflected the fact that at least a part of my income was a result of documenting other people's knowledge and that a share should go to them. But of course, there was no such thing. It appeared to me that I was also an exploiter. The moneylenders exploited the money market, landlords exploited the labour market, the traders exploited the commodity market, but I was an intellectual and I was exploiting the ideas market. I was taking the

knowledge of people, writing about it, but very little of that went back to them because I wrote mostly in English, and people as you know, don't understand the language in villages. A soul-searching began. It occurred to me that while my dilemma was not unique, the solution would have to be.

One day a thought came to my mind—honeybees do what intellectuals don't do. They collect nectar from flowers, but the flowers don't complain. In fact, flowers attract the honeybee. The bee not only connects one flower to another flower, but also does not keep the honey for itself. I realised that if I could share what I did with the people in their language, that is, give them credit and don't keep them anonymous, if I helped them learn from each other and if I did extract any rent from this effort, and if a reasonable share went back to them, then I could be like a honeybee. That would be an ethical, authentic way of living.

True to its name the Honey Bee Network implies basically four principles—cross pollination of ideas in local languages, acknowledgement of individual and community creativity without making them anonymous, protecting their knowledge rights, and sharing the benefits in a fair and just manner accrued from value addition in the innovations or traditional knowledge. Accordingly, about 1,40,000 ideas, innovations and traditional knowledge practices (not all unique) have been mobilised from 545 districts of India. More than 90 per cent have been collected by the volunteers of Honey Bee Network while the remaining have come in response to the advertisements.

The Honey Bee Network was started in 1988-89 leading to the establishment of Society for Research and Initiatives for Sustainable Technologies and Institutions (SRISTI) in 1993. Then in 1997 as a follow up to an international conference on creativity and innovations at grass-roots at the Indian Institute of Management (IIM), Ahmedabad, the Grassroots Innovation Augmentation Network (GIAN) was set up in collaboration with the Gujarat government. The purpose of GIAN was to reduce transaction costs of innovators, investors and entrepreneurs by linking them with each other. It was also the first attempt to provide micro venture capital support to grass-roots innovations. The NIF was established and it has become a grant-in-aid institution under the Department of Science and Technology, Government of India.

The world is just a click away. Excited and curious children in a village in Champaran, Bihar. These are some of the many expressions that SRISTI and NIF have created in the country by enabling hundreds of people to realise their dreams.

SRISTI's suggestion led to the establishment of the Micro Venture Innovation Fund with the help of a banking institution. It has invested more than ₹2.5 crores in ideas and innovations by common people without collateral or guarantors. More than 60 per cent of the borrowers have paid back. Some sixty technology licenses have been given mainly to small companies and individual entrepreneurs with the benefits going back to the innovators. In about two dozen cases, small entrepreneurs chose to get technologies after paying money even when patents had not been granted. Their ethical standards were admirable. If they had just copied and commercialised the innovation, no legal action could have been taken against them. But they did not deprive the innovator of his due. Probably the values of the Honey Bee Network had influenced them.

SRISTI has been organising biannual *shodh yatra*s for the last thirteen years. The idea is to honour knowledge experts. The message is that traditional knowledge as well as contemporary innovations matter. Visitors from far-off places coming and honouring local institutions and individuals send a signal to the local communities. The innovators are then seen with some respect in their villages. It is part of the institution-building effort. India's knowledge society cannot evolve unless the worth of local knowledge is recognised. To promote the spirit of excellence and collegiality in villages, biodiversity, recipe and idea competitions are also organised. The biodiversity competitions provide a way of speeding up the knowledge transfer from grandparents to the grandchildren. Through open book quizzes, children can learn from anybody in the village about plants and their uses and bring their collection to the meeting. Sometimes the children who excel in this domain may not be very good in studies. The cultural creativity is also celebrated with a view to finding markets for products through E-commerce and other channels.

Some of the dishes cooked in villages use ingredients that were seen as weeds growing naturally and special attention is given to these recipes. The changing climate may endanger some of the present agricultural products and thus new food sources will be useful. Many of the so-called weeds are rich source of nutrition. The survival instinct and the inquisitiveness of the poor people might actually hold the key to the survival of humanity. Thus attention to their knowledge is also vital because that will provide ways of survival for the more privileged ones who have lost such an instinct. We also take the blessings of centenarians on the way and try to learn from their life experience. Many times innovators are discovered serendipitously.

The two cases described in the accompanying pieces represent a very small sample of grass-roots innovations. These are a pressure cooker-based coffee-making machine by a roadside mechanic and a baffle septic tank system installed in a PVC tank by a small mechanic in Kerala. These represent different dimensions of grass-roots innovations system in the country.

Another story of Mansukhbhai's motorcycle-based ploughing machine demonstrates how a shared problem in a specific socio-ecological context could trigger an innovation that is copied, adapted, improved by a large number of other mechanics to meet distributed need. A very vibrant informal innovation ecosystem is evident from this case.

Many of the innovative products have been sold in various countries around the world amplifying what we call as G2G—grass-roots to global model. The creation of a public good alongside licensing of technologies represent a portfolio of options some determined by the innovators and some by the market. The government has played an active role in recognising the common creative people. The experience of the Honey Bee Network carries important lessons in inclusive innovation and economic development. Recently, the American Society of Mechanical Engineering (ASME) through its Engineers for Change programme has sought partnership with Honey Bee Network. Scientists Beyond Borders has sought a similar partnership. A global innovation foundation based on the philosophy and practice of Honey Bee Network may not be too far. If creative people with opportunities to craft their own ecosystems come together, it may also help to make the world more compassionate and collaborative.

*—Anil K Gupta is the Founder,
Honey Bee Network and Executive
Vice Chairman, National Innovation Foundation.*

Once an innovator, always an innovator

TR Rajesh of village Pannacherry in Kerala is a young man who has wandered all over the country and worked in a lathe workshop and on a construction site and did several other odd jobs. He finished school education up to standard ninth but with a large family to support had to start earning at an early age. At the age of seventeen, he left for Mumbai in search of job. After being a welder, lathe operator, die-caster, electrician, plumber, driver, draftsman, mason, etc., he learnt to take life with equanimity and enormous curiosity. However, he decided to return to Kerala when he was stricken by jaundice.

Once back to his home state, he started taking up contract for constructing houses. During that time he began working towards bringing down the cost of a septic tank as that would help him in cutting down the total cost of construction. He borrowed books on engineering from a public library and started reading on the subject. Within two weeks, he developed the prototype of a low-cost septic tank and installed it in his neighbour's house. The system worked very well for five years. However, the local research institute would not even test his idea but his desire to perfect the solution led him to spot a newspaper advertisement given by the NIF inviting innovators to present their work. NIF helped him to file a patent on his behalf in 2004. Recently, a technology for which he got a national award was licensed to a company in Goa. The award followed by the license made him more determined to continue innovative work even if it does not get much money. He began to help mechanics with their problems. He sold vacuum cleaners and moulds for rubber retreading. He came out with his second innovation 'Biokitch' which helps in digesting kitchen waste to generate biogas. A group of entrepreneurs set up a company to commercialise his technology and have given him 10 per cent stake. He has been appointed as Chief Innovation Officer of the company. He wondered how to design a valve which would allow the waste to flow while stopping the release of the smell. Once while travelling in a train, he thought of a 'ball valve' which solved his problem. He received ₹1 lakh as an advance royalty and more substantial regular earnings were assured.

He is very methodical in his approach to innovations. He is skilled in making moulds and therefore knows the importance of technical drawing. He has some more innovations waiting to be turned into prototypes. In the true spirit of the Honey Bee Network this innovator finds time to look for and help other innovators. He feels truly delighted at the success of his fellow innovator Augustine Thomson, whom he had mentored. Augustine won the Kerala state award for his Electro Tyre Retreading machine, which ensures cost-effective tyre retreading process.

TR Rajesh advises every innovator to be critical of his own ideas and take a detached view while working on these.

—TJ James

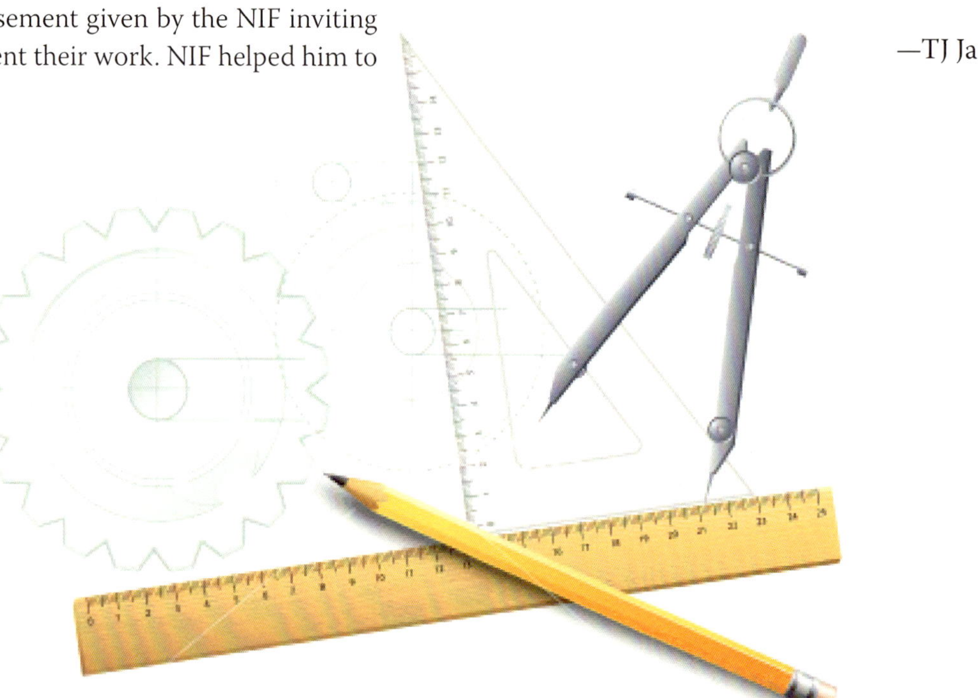

Rozadeen's coffee machine for chai wallah

Almost every tea stall in the small town of Motihari, famous for Gandhiji's successful experiment of satyagraha against the exploitation of indigo farmers, flaunts a contraption named 'JP Ustad Coffee Cooker' developed by Mohammad Rozadeen alias JP Ustad. The forty-seven-year-old gas welder modified a pressure cooker to incorporate a copper delivery pipe on its lid to transfer the steam generated inside to a container outside. In the process, he converted an ordinary cooker into a coffee-making machine.

A keen sense of observation led to the development of this device. At every marriage party, Rozadeen used to observe people thronging at the coffee stall. He reflected and realised that not many actually drink coffee otherwise in their daily lives and prefer tea. So what makes them drink coffee at these marriage parties? And if they liked coffee so much why are they not starting to make it at home? He realised that coffee prepared like tea was not acceptable to most as they preferred coffee to be hot, frothy and espresso-like. And to have such a cup, nobody was willing to shell out as much as ₹3,000 for home coffee machines. The other part of the problem was that coffee was not an available option at all places and most tea stalls did not serve coffee.

Rozadeen pondered if he could make a low-cost device not requiring electricity, it may help him earn some money and supplement the income of tea vendors as well. He conceived this idea but it involved some investment and he came from a very poor background with his wife, four daughters and two sons to support. An investment of a few hundred rupees from his already paltry income without the guarantee of success was a risky proposition. Treading cautiously he made a prototype using an old cooker and tubes of copper. For this he had to take a loan of ₹2,000 from one of his friends, which he repaid later.

The working of the device is quite simple. Water is heated in the cooker. Once sufficient steam pressure has been obtained, the steam is led out by moving the lever attached to the copper delivery pipe fitted on the lid. The delivery tip of the copper pipe has been constricted to create more pressure at the point of release of the steam.

The cost of 'coffee-cooker' varies as per the size and requirements. A new 10 litre capacity cooker costs about ₹2,000 while a modified old one costs ₹750. Similarly a new 5 litre cooker costs ₹1,000 and an old one ₹500. If someone brings his own cooker, he does the modifications for ₹250. The NIF has filed a patent in his name for the coffee cooker and conferred on him an award and felicitated during the twenty-second *shodh yatra* in Bihar in 2008-09. Rozadeen has sold over 1,500 such 'coffee-cookers' in Motihari and several other nearby towns.

His machine has drawn a good feedback from people who have switched to coffee from tea and the demand picks in the winter season. The vendors sell a cup of coffee between ₹6-10 and some customers have even appreciated the taste of the beverage—it tastes unlike the one they had sipped earlier at marriage parties. Feeling delighted at the success of his cooker, Rozadeen at present makes them only on demand, but hopes that some day in the future he will have a strong financial support, with which it will be easy to cater to the demands for the product as he points out, *'Paisa hoi tabhina dimaag badhi'* (my mind will work, only when I will have money).

—Nitin Maurya, Manoranjan and
Azhar Hussain Ansari

sharing the INDIA idea

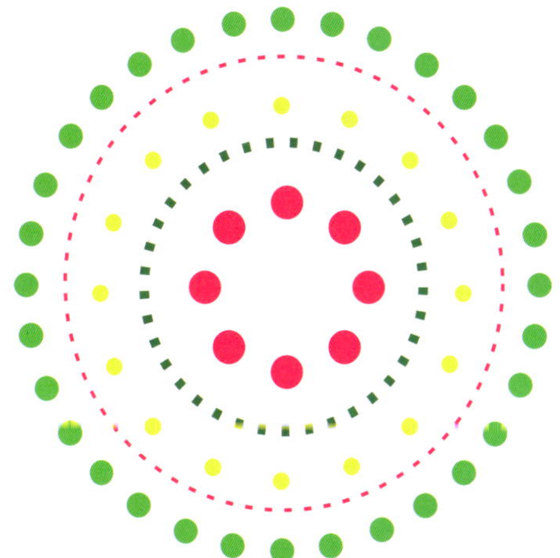

Jaipur Foot—
Transforming Lives through a Prosthetic Marvel

The sweetest sound they know of is of their footsteps. From Afghanistan to Rwanda, from Iraq to Vietnam and Pakistan to Sri Lanka...many of those disabled by wars share a common bond, a special affiliation and a moving saga of despair and joy. They and thousands of others in different parts of the world, including India, have been able to discard their crutches and calipers and once again walk on their own. Their gloomy faces are lit up with hope; a cheerful smile speaks of their happiness and confidence. All thanks to Jaipur Foot, the world-famous artificial limb that draws its name from the city where it was developed.

Hari Singh from Alwar, Rajasthan, lost all hope in 1990 when a train accident shattered his dreams and chances of earning a decent livelihood. He says, '*Zindagi char aana rah gayee thi, ab baarah aana vapas mil gayee hai*' (Three-fourths of his life was gone but now he has recovered most of it). A vendor at the railway station, he got his third Jaipur Foot this year since his first fitment eleven years ago. He gets overwhelmed recalling the dreadful accident and how he regained confidence with Jaipur Foot. He says he can't fly but he is able to earn his livelihood. Ram Behari from Banda, in Uttar Pradesh, is another amputee who gained confidence with the Jaipur Foot and is again able to plough his field and look after the family.

In the Jaipur Foot fitment camps in Kabul, Baghdad, Vavunia or Karachi, the patients celebrate a new life. They can rise on their feet again; they can run, pedal a bicycle, climb a tree, squat and even dance. The popular Bollywood movie *Nache Mayuri* (1986) recounts the moving tale of a budding Bharatanatyam

(page 80) Artists performing Bharatanatyam on the occasion of the millennium celebrations at the Raja Rajesvaram Temple, Tanjore, in Tamil Nadu. This all-stone temple was built by Raja Raja Chola, who ruled the region from 985-1014 AD. Indian classical dance-forms trace back to the *Natyashastra*, the ancient treatise written sometime during the period 200 BC and 200 AD by Bharata Muni.

(facing page top) Happy feet—a dance class in progress at a studio in Bengaluru. The Jaipur Foot has given a new lease of life to many individuals by helping them effectively deal with the loss of a limb and move on towards realising their dreams.

(facing page bottom) An Indian member of the UN peacekeeping force adjusts prosthetic limbs at a camp in the village of Ibl al-Saqi in south Lebanon. Lebanese victims of cluster bombs were provided artificial limbs at this camp.

dancer Sudha Chandran who lost her leg in an accident but started life anew with Jaipur Foot. There are people from all walks of life—farmers, soldiers, workers, sports persons and landmine victims—who have immensely benefited with this artificial limb. Vinod Rawat could take part in a half marathon in 2008 after he got this artificial limb. Many others go back to the fields, factories or shops wearing a Jaipur Foot.

Since its inception in 1968 as a result of the efforts of a master craftsman, Pandit Ram Chandra Sharma, fondly called 'Masterji', Jaipur Foot has walked a long way in its illustrious career. Masterji conceived the idea of designing such an artificial limb and developed it under the supervision of Dr Pramod Karan Sethi, a noted orthopedic surgeon. Dr Sethi got Ramon Magsaysay Award for the invention in 1981.

The international scientific community learnt of the innovation when Dr Sethi presented the foot before British orthopedic surgeons in Oxford in 1971. They applauded the visiting surgeon. Dr Sethi eventually left his job at the government hospital and moved to a private hospital in the same city and fitted the artificial limb to several amputees. Thus Jaipur Foot became synonymous with him.

Jaipur Foot Walks Abroad

But it was an NGO working for the disabled that spread the message about Jaipur Foot and transformed the lives of thousands of disabled people who were earlier unable to afford an artificial limb. Such people in nearly twenty-five countries know of the Bhagwan Mahaveer Viklang Sahayata Samiti (BMVSS). This voluntary organisation of Jaipur provides one stop facility, free lodging, boarding and fitment of Jaipur Foot to all who approach it. Led by its patron, DR Mehta, a former Deputy Governor of the Reserve Bank of India, this organisation has helped 3,051 people in Afghanistan, 3,000 in the Philippines, 1,800 in Sudan, 1,200 in Sri Lanka and 1,000 in Bangladesh.

Jaipur Foot—Then and Now

Initially made of aluminium with open ended socket, Jaipur Foot has been improved over the years with continuing R&D effort. The latest model of the limb is lighter, more functional and durable and in design and functioning closer to the human limb. Jaipur Foot consists of two parts—the foot piece and the socket—with or without the joint, depending on whether a below the knee or above the knee fitment

is required. Instead of wood or aluminium sockets, now High Density Polyethylene Pipe (HDPE) sockets are being made using a German technology. With HDPE, Jaipur limb has a more comfortable stump-socket interface. The external skin made of vulcanised rubber has added to the strength and range of movement. With these feet, the gait is so natural that even a farmer is able to walk on a rugged terrain with remarkable ease, says Mehta.

Jaipur Foot is almost a prosthetic marvel having all desirable human movements—dorsi-flexion, inversion, eversion, supination, pronation and axial rotation—which make squatting, climbing trees and sitting cross-legged possible. This waterproof limb lasts for about three to four years and is the most cost-effective artificial limb in the world. It costs only around US$45 while the cost of artificial limbs starts at US$12,000 across the world. More than 4,00,000 Jaipur Foot have been fitted since its creation and many more people have been given calipers, crutches, etc. Unlike Western Sach Foot that requires a closed shoe to protect and hide it, Jaipur Foot can be worn with or without shoes. So Jaipur Foot users have no problem visiting temples or mosques.

ISRO Foot and Jaipur Knee

Field trial for a polyurethane foot is underway to make Jaipur Foot cosmetically better and long lasting, Mehta says. It has been designed by an ISRO team. The Dow Chemicals of USA, Pinnacle Industries of Indore and Liberty Shoe Company of New Delhi are also collaborating in the trial.

The NGO has gone on to develop Jaipur Knee in collaboration with Armand Nukermans at the Stanford University. It has come as a boon for the poor patients who loose their knee joints. Made of self lubricating, oil filled nylon, pretty close to the human joint, Jaipur Knee (costing just US$20) got listed as *Time* magazine's fifty best inventions in 2009.

BMVSS, Camps and Training Programmes

At BMVSS, doctors treat all patients, even the poorest of the poor, with utmost respect. The organisation was set up on the occasion of 2,500th death anniversary of Lord Mahavir. DR Mehta took a vow to help the disabled after he recovered from a serious accident in which the bones of his left leg were almost crushed. His leg was not amputated but Mehta did not forget the agony of those who lose their limbs in mishaps, natural calamities or diseases.

A management and value system that recognised the needs of the patients in terms of affordability made a great difference. 'We were able to fit almost 10,000 limbs in next seven years.' At present, more than 20,000 limbs are fitted every year. 'Perhaps no other organisation fits more limbs in the world,' he says with a great sense of achievement.

BMVSS has twenty centres in India and many abroad including two in the Philippines, two in Pakistan and others in Columbia, Kenya, Uganda, Manila, Nairobi, Nigeria, Rwanda, Honduras, Panama, and Trinidad and Tobago. The organisation also trains local technicians at camp sites so that they can continue the work after BMVSS team returns to India. Wherever they go, Jaipur Foot team is greeted with great admiration and love, recalls Mehta. An old lady in Pakistan asked him to take all her money and use it for the charitable work by his organisation.

Jaipur Foot has also earned tremendous goodwill for India. Pakistan's newspaper *Dawn* wrote in 2002, 'As aid pours into Afghanistan, a special consignment from India is probably bringing more happiness to Kabul than the rest of the world's cargo combined. The consignment consists of 1,000 pieces of the Jaipur Foot, a prosthesis named after the north Indian city where it was developed in 1970. As goodwill gestures go, it has probably earned India more appreciation than any amount of diplomacy.'

Jaipur Foot indeed has come a long way. Yet it has miles to go so that many more people may walk.

—Abha Sharma

A patient tries to walk with an artificial limb at the BMVSS in Jaipur. To ensure that the needy get speedy solutions to their problems, the limbs are made on the spot and fitted. Only when the patient is satisfied with the new limb is he sent back home.

Malkha—
when Technology Meets Craft

'Our work began with an almost theological belief that God did not create the cotton bud hoping that man would make big machines.' This thought came to L Kannan, an IIT Madras alumnus, as he saw a technology gap affecting the traditional handloom weavers. At the same time, Uzramma, Founder of the Dastkar Andhra organisation, with her long experience of working in the handloom sector, wanted the weavers to be helped by a system that integrated cotton cultivation in the field with the spinning of the yarn in the village for supply to the loom in the cottage. Kannan and his team developed a micro spinning machine that helps in setting up a complete village-based cotton textile production chain—from cotton growing to spinning to hand weaving.

What comes forth is the fabric named Malkha! This unique fabric brings together machine made thread and handcrafted handloom cloth in a stunning reconciliation of two paradigms—the primacy of the village and the economy as well as aesthetics and enablement from the little machine. It makes it possible to revive a way of life, sustained by peasant-weaver communities for about 4,000 years and to realise Mahatma Gandhi's dream of self-reliant villages! In the human context of the new economics, work is diverse, home-based or local and everybody participates. Across India, there are some 6 million people still working on the loom, more than half of them being women. After agriculture, it is the second largest provider of employment. It produces around 700 million metres of handloom fabrics of distinctive varieties and unique qualities. But the yarn for the loom has to come from the mills far away. The spinning wheel is no longer very viable. The supplies of yarn from factories get increasingly erratic and uncertain. With this support dwindling, traditional skillful professions and families break up. The men, once proud skilled craftsmen, migrate to look for unskilled jobs in the cities. Unless the handloom industry is revived at the ground level, the loss will be beyond computing.

As in any breakthrough innovation, the gains are multifarious and unforeseen. A small locally situated machine would enable the handloom sector to have its own independent access to local yarn supply.

The bulk of yarn being produced in far-off factories is primarily sold to textile mills in cities which places handlooms in dire straits. Above all, Uzramma, Managing Trustee, Malkha Marketing Trust, reasons, 'In a world where the highest cost in the future is going to be that of fuel, such a local model is not only relevant for the manufacture-consumption of today, but is also the future of manufacturers.' The decentralised cotton processing is energy efficient, eco-friendly and cost-effective as it uses locally grown cotton to weave fabric and in the process cuts out on the pollution caused by transportation of the raw material from the village to the factory. It also promotes biodiversity to which India's small cotton farmers are used to.

The process of baling cotton—compressing it into tight bales—was introduced to transport Indian raw cotton easily to the mills in Great Britain, under the colonial rule around the late nineteenth century. The micro spinning machine not only handles cotton gently but adjusts with ease to any of the different local varieties some of which are found unsuitable

by big spinning mills that treat all cotton in the same manner. The village-based system desists from torturing and compressing the freshly plucked cotton into tight bales, to be loaded and carried away to a distant destination where it is made to pass through a mechanical process to make it fit for fast spinning machines.

Kannan points out that the conventional cotton spinning mills involve very huge machines that tear into the fibres and organise them in a manner that makes them amenable to yarn spinning. 'Our key conceptual insight was that a lot of the operations could be avoided or simplified, if the act of pressing cotton into bales was avoided and I saw no rationale to persist with this if spinning and cotton growing were to be co-located.'

Apart from the aesthetic revulsion for this method of yarn production, this was a bottle-neck in the creation of an inclusive textile industry. Kannan says, 'Fortunately, I had no prior background in textiles and was therefore unfazed by all the "experts" who were unequivocal in their opinion that no small-scale technology was going to be feasible. But surprisingly, I learnt later that mine was not an entirely original insight—there is an industry proverb

Dexterous hands at work. Women form an important part of the workforce at the Malkha production centres, which are spread across South India.

which says "when the cotton is in the bud that is as good as it ever gets". This is used to tell engineers that all machines damage cotton, only, better machines inflict less damage. So I decided to build a machine that would "do" very little to the cotton and instead organise the fibres largely through aerodynamics and buoyancy. That would lead to better quality as well as reduced cost of machinery, I reasoned.'

The decentralised cotton processing reduces the costs and requires radically less power and water. The micro spinning machine has safety features against power fluctuations and inbuilt diagnostic tools for trouble-shooting and easier maintenance. Each unit can support the work of around 100 families. There are various Malkha producing centres working in Andhra Pradesh, Karnataka, Maharashtra and Kerala. The Kerala unit is run entirely by women. The integrated system of cotton growing, spinning and weaving allows small producers to take charge of production and management. The NGOs and trusts help them manage the units and also assist them with promotion and marketing.

Kannan's team worked with the support of CSIR, Anna University and other agencies. The main innovation in this cotton processing system has been in developing a machine which 'plays' with the natural buoyancy of cotton, touching the fibre lightly to let it regain its springiness. The social and economic gains from a village-based textile chain cannot be touched and felt but every fashion designer who has felt the texture of Malkha has been impressed by its unique quality. This handwoven pure cotton fabric has a beautiful texture, it is soft, absorbent, and drapes beautifully. The colours used for Malkha are either natural or completely non-toxic. Blue comes from natural indigo, yellow from pomegranate rind mixed with *harda*, *kasimi* is used for grey-green and brown comes from

(top) The cotton seed matures into a ball of creamy-white fibres. These are loose and have a natural twist, which makes the process of spinning the yarn easy.

(middle) The colourful Malkha fabric has attracted a lot of attention amongst fashion designers, both at home and abroad, because of the human touch which is involved in its production process. The workers delicately handle the fabric, thus providing it with a unique texture where every strand is visible.

(bottom) The Malkha fabric is best worn during extremely hot and humid conditions in India, as the simple, soft, hand-woven garment can breathe and does not stick to the body.

catechu heartwood. Even the reds come from the non-toxic alizarin.

Malkha has begun to command an international appeal. Inspired by the softness of the yarn and the natural feel of the fabric, designers use it for home and fashion products. The fabric is especially suitable for artisanal weaving. The decentralised production process meets with ease customised small orders coming from diverse customers. Uzramma says this technology—given its ability to expand to larger scales—will one day be able to cater to local rural markets as well. The inbuilt flexibility of decentralised manufacture will prove itself relevant in meeting the varied requirements of the handloom sector—from small and specialised, to large and of standard quality.

Uzramma points out, 'The strength of traditional pre-colonial Indian cotton textile technology was not just in the ability to provide sophisticated products for export, like the finest of cottons called muslin or *malmal* but also to supply ordinary cotton cloth on a vast scale, which lent itself to versatile uses. Such large-scale production was possible because the operations were enmeshed in society in a decentralised way. Production was home-based in villages; yarn was spun at home for domestic use; and in many cases, cotton was grown in the courtyards thus providing a year-round supply of raw material.'

Mahatma Gandhi picked up this very aspect of cotton production when he proposed khadi as a self-sustaining economic activity to counter India's impoverishment under colonial rule. Given the continuing relevance of handlooms to provide livelihood to millions, Malkha—drawn from the words *malmal* and khadi—therefore truly positions itself as the fabric of a new age.

—Mayank Mansingh Kaul

(facing page top) Thousands of men and women work hard at the Malkha production centres to cater to the worldwide demands for this sophisticated and elegant fabric.

(facing page bottom) A woman working at a Malkha production centre. In 2007, the Ushodaya Grama Samakhya organisation from Punukula village in Andhra Pradesh opted for the making of Malkha cloth, giving employment to twenty-four families at the training stage and producing 215 metres of fine fabric.

(top and bottom) The fashion industry in India is attempting to work in tandem with the craftsmen at the grass-roots level.

A **Potter's** Tale

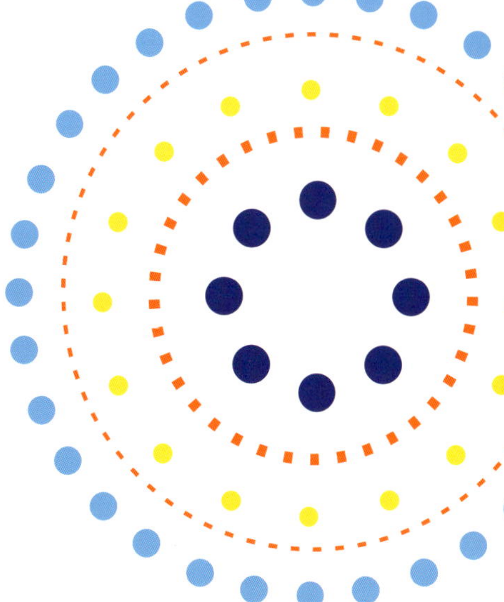

The potter does indeed know his clay. The silent dialogue between man and mud continues down the generations. 'I am a potter. My father made earthen pots for the villagers who paid him in grains…' Mansukhbhai Prajapati, a school dropout, began to speak in hesitant Hindi in the Siri Fort Auditorium of New Delhi. Mansukhbhai was invited to describe his entrepreneurial journey and to exhibit his prime product—a refrigerator made of clay, sold under the name of Mitticool (mud cool).

More than a thousand people listening to him included technologists, venture capitalists, CEOs and young aspirants. TiE, the global organisation of Indus Entrepreneurs, had invited Mansukhbhai to honour this grass-roots innovator. TiE mentors young entrepreneurs who observed this serial innovator who had pursued his dream project without a business plan. Undaunted by the summit's elite audience—the potter from Nichimandal village in Rajkot, Gujarat—packed the story of his innovations with earthy humour and more punch than a PowerPoint presentation. The audience clapped every time Mansukhbhai recalled how he was able to overcome a particular hurdle that threatened to derail his product development work.

Loud applause greeted Mansukhbhai again when his clay fridge was brought to the stage. This was his showpiece, but was only one of several innovative products in clay. The fridge costs about ₹2,000 and uses no electricity. The cooling is done through evaporation. The clay is baked at high temperatures and hence the handling of the fridge does not call for special care. As a lad of eighteen, Mansukhbhai had devised and built a clay mixing machine for the potters. This fridge reflected a combination of traditional knowledge and technology inputs, a modern contraption painted with a multicoloured ethnic design. The fridge can preserve fruits and vegetables for about five days and keeps milk from getting spoilt up to three days. Mansukhbhai continues to make further improvements, though these make it a bit more expensive. With the help of the NIF and the National Institute of Design (NID), he has come out with a model with two tiny fans that run on solar power. More than 1,000 refrigerators have been sold but Mansukhbhai says he gets more orders from the urban areas than from villages, which was not the idea he started out with. His creation was primarily

(facing page) Potters at their workshops with newly created earthenware. The origins of pottery can be traced back to the pre-Harappan era, when clay was available in abundance around the riverbeds, which were the hubs of ancient civilisations.

meant for people living in villages who have zero or limited access to electricity. Mansukhbhai has no sales force but customers find him through newspapers, magazines and websites, etc. that mention his innovations regularly. His products have also reached foreign shores—London, Africa, Singapore and America. His organisation, Mitticool Clay Creation has sold nearly 1,00,000 products till date, fetching revenues around ₹30 lakhs.

How did Mansukhbhai think of a low-cost fridge? The earthquake of January 2001, caused havoc and also destroyed everything in the vicinity from buildings to simple earthen pots. A Gujarati

(left) A Mandna painting depicting a bird at Jawahar Kala Kendra, Jaipur. Painted by the women of the region who draw nature-based motifs to decorate their homes, this art form is an expression of their experiences.

(right) A 'true scientist' for the former President of India, Dr APJ Abdul Kalam—Mansukhbhai's clay refrigerator has won him accolades both in India and abroad. The fridge has the capacity of 50 litres with another striking feature— the water which is used as a coolant can be consumed by opening an attached tap.

(facing page) Mansukhbhai's products are mainly targeted at rural consumers. In his zeal to create a simple and affordable product, in 2009 he made a clay pressure cooker which not only cooks delicious and healthy food, but is also environment-friendly.

newspaper published a photo feature on the earthquake and the caption of a picture showing a shattered pot read: 'The broken fridge of the poor.' The words 'Poor Man's Fridge' struck Mansukhbhai and ignited the idea he was looking for! The idea of an affordable rural refrigerator. He would try and make a fridge that can be bought and run by a poor family. He worked on a prototype for three years. While he was clear about using evaporation for cooling, it took a lot of time and effort to perfect the model. The final product was accepted by the customers who have not come back to him with any complaints.

Amid another round of applause, Mansukhbhai went on to show a water filter, a pressure cooker and a non-stick tawa—all made of clay. The water filter costing less than ₹400 caught the eye of a trader who then exported some pieces to Nairobi. These products and Mansukhbhai's recognition as an innovator by the President of India, all came as fruit of the potter's zeal and the determination to achieve something, to do something for ordinary folk like him: *Mujhe kuuch karna hai, aage badhna hai.* (I must do something. Must move ahead in life.)

The family needed money to survive. Mansukhbhai had not passed even the higher secondary examination and thus did not have many career options. His first break came when he got a job as a helper at a highway teashop. But, 'I was engaged to be married and unfortunately, my prospective parents-in-law would pass by and see me working, much to my embarrassment.' The proud potter, a craftsperson, did not quite see himself serving tea and washing up rest of his life.

His release from that teashop came when he was offered a job in a roof tiles factory nearby. He was at home once again, dealing with clay and sand and imbibing the new techniques being used on the material so familiar to him and his forefathers. His work won the employer's praise but Mansukhbhai was yearning to do something on his own. He decided to take a big risk and asked a moneylender for a loan of ₹50,000. His employer was ready to vouch for his talent, but when the moneylender bumped into Mansukhbhai's father, the risky deal nearly fell through! It was the moneylender who counselled the reluctant father to invest in the ambition of his talented son. The moneylender was an exception because most of the villagers had not thought much of the lad's capability. The loan amount was brought down to ₹30,000 to reduce the risk. The father still wondered whether it was wise for his son to leave a job that gave him ₹300 per month just to chase some dream that may lead to disaster. But while Mansukhbhai got the money for his experiments, a long struggle ensued before he got the product right. Trial and error marks all development. Mansukhbhai gave the example of his clay tawa, a substitute for the iron one. He had to discard some 1,00,000 pieces before he got this simple product right.

Mansukhbhai's constant endeavour to develop new inexpensive products gave him a lot of personal satisfaction but his problems as a budding entrepreneur with limited resources continued to multiply. He carried on without hope till one day; he was approached by Prof Anil Gupta, Executive Vice Chairman, NIF. NIF had been scouting for grass-roots innovators for some years and they came to hear of Mansukhbhai's work. Their timely financial help saved the project and Mansukhbhai perfected his Mitticool by 2005. Now, Mansukhbhai's son is doing ceramic engineering and the villagers respect him. After all, no one else in the village has been given an award by the President of India or appeared in newspapers and on the TV! The fellow potters tell him, 'You are the pride of our community'. Prof Anil Gupta says Mansukhbhai's enterprise, 'Combines technology with traditional knowledge to deliver sustainable solutions.'

Mansukhbhai had not heard of Prof CK Prahlad, the visionary who had coined the term 'Bottom of the Pyramid', but he himself lived a life of the marginalised. Each of his products was inspired by his wanting to do something for those like him. The thought of villagers drinking unsafe drinking water and being stricken with diseases made him develop an earthen water filter. When his wife asked him to buy a non-stick tawa, he realised that a poor man cannot afford it. That inspired him to make one out of clay. Mansukhbhai believes that the use of earthen ware for cooking and eating is good for health. He considered aluminium utensils to be undesirable and went on to perfect a pressure cooker made of clay. It was held up before the large audience. But that was not enough. Mansukhbhai opened the cooker's lid and blew into it from his mouth. The loud sound of the whistle amused the audience!

Mansukhbhai is not resting on his laurels. At the TiE summit, he ends his narration by talking of making a 'Mittihouse', mud house that provides protection from extreme temperatures and needs no artificial light. Hearing the last words of Mansukhbhai, the next speaker at the summit panicked and urged him to delay his killer innovation—this technocrat was to begin his presentation on a solar lamp!

—LK Sharma

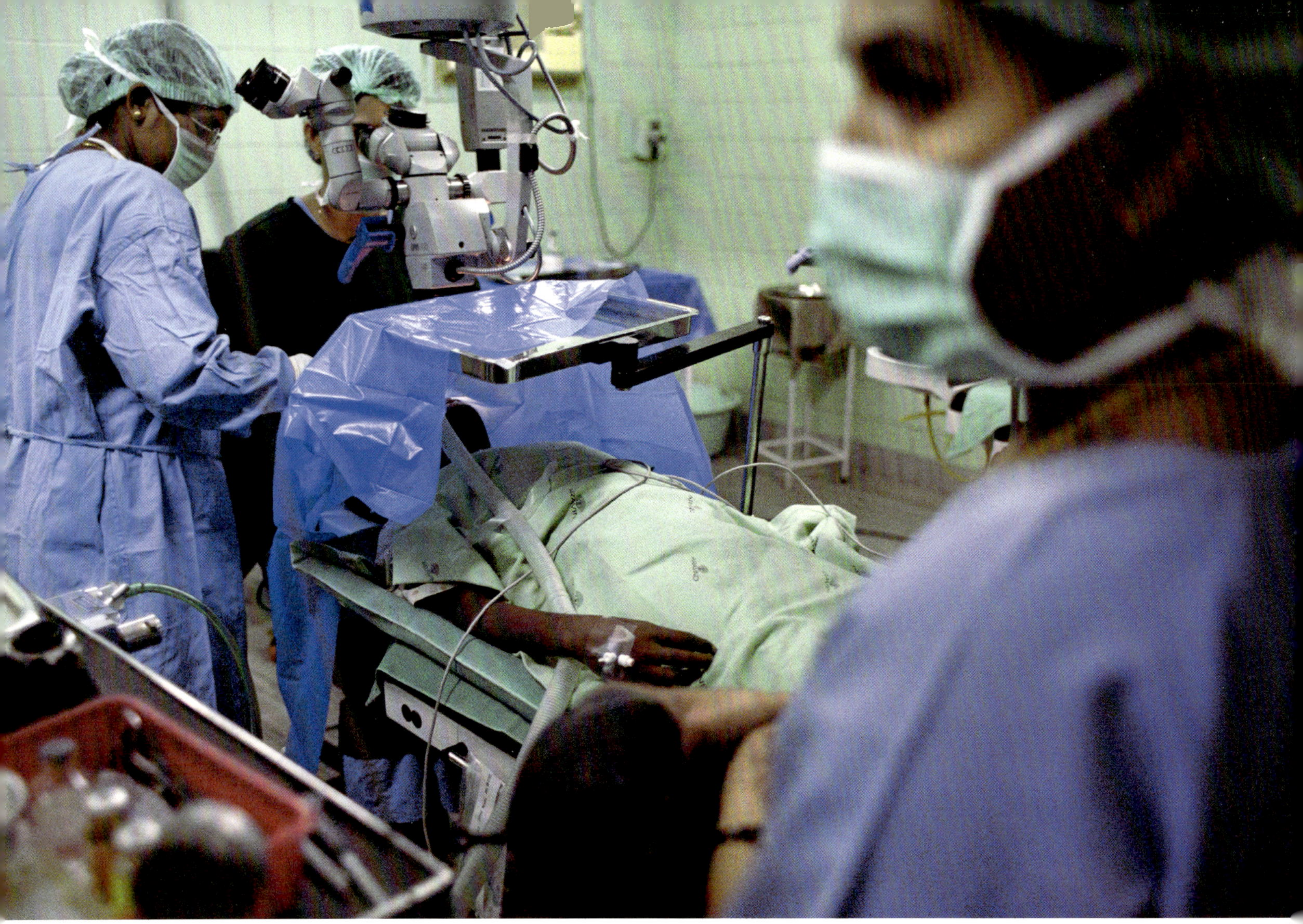

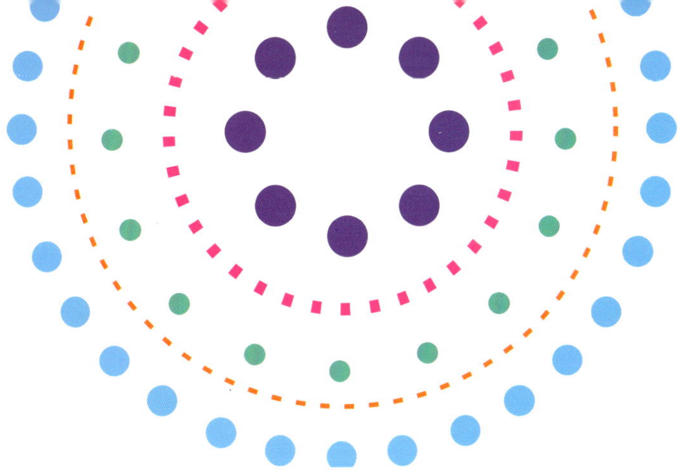

Aravind Eye Care—
the Vison Visionaries

'The city of Madurai is best known for its spectacular Hindu temples. Built several years ago, they still attract thousands of worshippers. But in recent years, Madurai has been attracting thousands of new visitors. These are not sightseers, rather they are people seeking to have their sight restored—pilgrims of modern medicine. They come to Aravind Eye Hospital at Madurai which many experts consider one of the finest eye hospitals in the world.'

—*The MacNeil/Lehrer NewsHour*, 1989

Avoidable blindness among the poor always perturbed eye surgeon Dr Govindappa Venkataswamy. On retiring from government service in 1976, he decided to continue his mission and established an eleven-bed clinic with the help of his sister and brother-in-law, also ophthalmologists. Today the multi-facility, Aravind Eye Care System treats more than 2 million patients a year. At multiple locations, a quarter of a million surgical operations are performed every year. The organisation provides 45 per cent of eye care in the state of Tamil Nadu. This has been made possible through a business model innovation.

The first barrier to good eye care in a country like India was letting people know that treatment was available. Millions of people in India suffer from full or partial blindness, which in about 80 per cent cases is caused by easily treatable cataracts. Those affected, especially among the rural poor, often assume either that there is no treatment, or that it would be unaffordable. Aravind Eye Care System creates a demand, an understanding, that if you have a cataract, a simple operation can fix it and you can again support your family.

Through a network of five eye hospitals and screening camps, Aravind provides comprehensive eye examinations, advanced state-of-the-art treatment and rehabilitation. Aravind's unique model of combining high volume with the highest quality standards enables it to provide affordable treatment.

(facing page top) Surgeons at work at the Aravind Eye Hospital in Madurai. Named after the great spiritual leader, Sri Aurobindo, the organisation has set up exceptional benchmarks in eye care. In recognition of its endeavour to reach those who need its services the most, in 2010 Aravind Eye Care System won the world's prestigious Conrad N Hilton Humanitarian Prize.

(facing page bottom) Manufacturing of lenses and blades in progress at a sterilised set-up at Aurolab. Established in the nineties, it has made Aravind Eye Care System self-reliant and self-sufficient by providing high quality products to the organisation.

The eye care provided in Tamil Nadu has contributed to a decreased prevalence of blindness in the state. Aravind performs about 5 per cent of the ophthalmic procedures nationwide, with less than 1 per cent of India's ophthalmic manpower. In 2009-10, 2.5 million people received eye care at Aravind and 3,00,000 surgeries were performed, with 70 per cent of the patients receiving free or subsidised treatment. The system's outreach programme generated 40 per cent of those surgeries. This programme encourages active involvement of the community. Working with a large community network, Aravind conducts over 1,300 eye camps annually. Aravind's outreach model goes beyond the traditional cataract-only screening. It provides comprehensive eye examinations and necessary treatment at the base hospitals. It is the single largest eye cataract surgery provider under one roof in the world.

Vision Centres

Aravind has also established a growing network of 'Vision Centres' with easy access for the rural population. These centres have data connectivity to the base hospitals employing low-cost technology to provide real time consultations. As of October 2010, 3,50,000 patients have received consultation at thirty-six centres. Each month, about 70 per cent of the patients visiting the vision centres are first-time patients, while the remaining 30 per cent come in for follow-ups. These permanent vision centres are significant for two reasons. First, this implies much better penetration of eye care than the eye camp approach since steady streams of new patients get treatment every month compared to an occasional burst of patients at eye camps. Second, it also implies a higher quality of eye care as seen by the high ratio of follow-ups. Patient interviews from these areas indicate that but for the tele-ophthalmology facility at the centre, they would not have received treatment.

These rural vision centres are staffed with trained ophthalmic technicians who carry out all the basic diagnostic procedures including refraction and examination. Patients, who require glasses, have immediate access to them, as the technician is also trained in dispensing spectacles. These rural centres have low-cost broadband connectivity to interact with the base hospital. All cases are discussed with the ophthalmologist at the base hospital via real time videoconferencing. Not a single patient is given any prescription without his authorisation.

Only those requiring further examination or surgery, (roughly about 10 per cent) are referred to the hospital.

These centres are financially sustainable due to the low operating expense of the network. Eventually, all eye camps will be phased out and replaced by permanent vision centres offering low-cost tele-ophthalmology. The vision centres' contribution to the target population is evident through the following indicators:

- Better health seeking behaviour in the community: Through the vision centres, Aravind has reached and served nearly 80 per cent of those in need of eye care within a period of two years in Tamil Nadu, the seventh most populous state in India. This shows the positive change in the attitude regarding health care, as patients who previously would wait for a camp, are now approaching the vision centre for their eye care needs.
- Comprehensive eye care to the community: Conventional eye camps are effective in identifying patients who need cataract surgery, whereas screening patients for other diseases such as diabetic retinopathy, glaucoma and others is far more difficult. However, Aravind's Vision Centres are equipped to diagnose and treat these diseases and ensure compliance to treatment—a critical aspect in managing chronic, progressive conditions.
- Reduction in the health care expenditure: The vision centres are equipped to provide comprehensive care to the majority of the patients; only 7 to 10 per cent of the patients are referred to base hospital for further treatment. The total cost of accessing eye care is now considerably lower for 90 per cent of the vision centre patients, as the incidental costs which includes travel, loss of wages and stay at the base hospital are eliminated.

Diabetic Retinopathy in VISION 2020 Priorities
VISION 2020—The Right to Sight, a global initiative launched by WHO has prioritised five key problems like cataract, refractive errors and low vision, trachoma, onchocerciasis and childhood blindness as priority areas. India has included diabetic retinopathy as one of the priority areas because the country has world's largest population of diabetics in the world. According to WHO figures of 2007, 40.9 million people were affected by diabetes in India and every diabetic is a potential candidate for

loss of vision due to diabetic retinopathy. It is well-known that only about 17 to 20 per cent of diabetic people will have diabetic retinopathy and need active intervention by the trained retinal surgeons and the rest of them will have normal fundus with good vision. They do not require retinopathy and need only a periodical follow-up by the diabetologists and ophthalmologists.

Mobile Screening with Use of IT

It is a known fact that there are not enough ophthalmologists to examine and screen all diabetic patients for diabetic retinopathy. Aravind Eye Care System has set up a mobile telemedicine facility that enables early detection of this disease or other blinding eye condition in diabetics by deploying qualified technicians at the screening level to capture high-quality images. In this model, a mobile van travels to rural areas or to physicians' offices where patients diagnosed with diabetes at that site are screened in the mobile van. The mobile van is equipped with a non-mydriatic camera to capture fundus images. This equipment is connected to a computer and to the videoconferencing unit. Images captured are electronically transferred to a reading and grading centre located at the base hospital. From the grader's input for each image, the software automatically elicits the severity level along with advice for treatment in a report format.

This information is relayed back immediately to the camp site where the report is printed and given to the patient who then receives counselling based upon the report.

Thus it was an appropriate time to concentrate on diabetic retinopathy and bring the problem under control as was done in the case of cataract blindness. However, the issues in managing vision impairment due to diabetic retinopathy are different from cataract. While cataract blindness is curable by a simple one time surgical intervention, diabetic retinopathy encompasses a multitude of problems and can be prevented if detected early and treated. It is an asymptomatic condition at the treatable stage but when a person presents for treatment with loss of vision, it often is too late for intervention. Aravind has developed a comprehensive eye care strategy to tackle this problem.

Research

The research activities at Aravind began within a year of its establishment in 1976, leading to one of the first publications documenting barriers to accessing eye care. Since then, operations research and research in basic sciences and clinical practice have provided inputs for improving care at Aravind and elsewhere and have influenced policy. The Dr G Venkataswamy Eye Research Institute (GVERI) with its state-of-the-art infrastructure facility has established to make

significant contributions in understanding the basic biological mechanisms of eye disease, clinical and operational research. It is affiliated with leading national and international institutes.

Aravind helped Seva Foundation to set up the Lumbini Eye Institute in Nepal. Dr Larry Brilliant, the Founder Director of Seva Foundation quotes:

'One of Seva's major projects, the Lumbini Eye Institute in Nepal, has grown from a small abandoned ward of a rural district hospital to its current campus with 220 beds and comprehensive eye care services. Lumbini now performs 25 per cent of all the sight restoring surgeries in Nepal. The Aravind approach to cost recovery was successfully adapted here for the first time outside India, as was its emphasis on reaching the entire community, regardless of patient's ability to pay for services. As a new generation of eye care providers emerged, Lumbini became a training and research centre, modelling itself after Aravind. Today, staff from China, Cambodia and Bangladesh comes to Lumbini for training just as decades earlier; the Lumbini staff went to Aravind.'

Aravind also helped a similar model at the Vivekananda Mission Ashram in Purba Medinipur, West Bengal.

Aurolab—a Manufacturing Unit

The Intraocular Lens (IOL) implanted during cataract surgery improves visual outcomes and, thereby, quality of life. However, the high cost threatened Aravind's ability to provide IOLs to poorer patients. Realising the importance of bringing down the cost of these lenses, Aravind established Aurolab, a fully developed manufacturing facility, which now produces high quality IOLs, sutures, blades and pharmaceuticals at a fraction of their cost in the West, enabling Aravind to maintain its equity in care. This benefit is now enjoyed by eye care providers and patients across the developing world.

In 1992, Aurolab introduced IOLs at US$10 while others were selling them at US$60-US$100. As Aurolab's sales grew, prices of IOLs fell worldwide, making Western companies sensitive to needs of the developing world. Aurolab now manufactures 1 million lenses annually and exports to 120 countries, with a global market share of 7 per cent. About 80 per cent of Aurolab's products go to various non-profit organisations.

Education and Training

In order to improve the medical and paramedical personnel resources, Aravind is providing the structure training for Mid-level Ophthalmic Personnel (MLIP). They constitute the backbone of the Aravind Eye Care System and enable it to provide

high quality and affordable services to a very large number of people. Aravind trains general ophthalmologists, highly skilled specialists, paramedic staff, hospital managers and a variety of support staff. Aravind's model of service and sustainability is influencing the thinking of the corporate world on reaching the poor and the lessons begin pretty early for some—the hospital is a crucial case study at Harvard Business School. Such has been the impressive achievements of an institution started by a philanthropist who believed that, 'Intelligence and capability are not enough. There must be the joy of doing something beautiful.'

—Dr P Namperumalsamy

Muruganantham's
Leap of Faith

In India, innovation may mean something completely different from just inventing machines, processes and protocols, as the story of A Muruganantham will tell you.

Today, his is well-known, at least in the IITs, IIMs and MIT, but five years ago, 'People ran away and hid themselves when they heard my moped arriving', he recalls.

For this forty-eight-years-old, Coimbatore-based social entrepreneur innovation has meant a long, long journey in completely uncharted sociological dimensions. Innovation has meant a huge battle with superstition and social taboos, ignorance, false data and even threat of divorce! Building his sanitary napkin-making machine has been the easier part of the job, he says, in a country of more than 1 billion where men are kept completely ignorant of the reproductive cycle.

Ten years ago, when he got married, Muruganantham was one such ignorant young man. Though he had grown up in a family of mother and sisters, he was blissfully kept unaware of the menstrual cycle that women go through and the associated need for hygiene, 'Like thousands of other men in this country', he says.

His father was a weaver and his mother a farm worker. After his father died, '₹7 daily was all we had for food and schooling', he recently told a web journal. A school dropout, he began working in a local workshop, 'earning ₹2 per week'. He was fascinated by all things mechanical, so he trained in small workshops and set up a grill-making workshop. When financial conditions permitted, he got married in 1997.

Innovation figures nowhere in Muruganantham's story so far, but in 2009 he was awarded the NIF's 'Fifth National Grassroots Technological Innovations and Traditional Knowledge Award' by President Pratibha Patil and the Massachusetts Institute of Technology bought Muruganantham's innovation—the 'low-cost mini sanitary napkin-making machine' for distribution in African countries.

Light dawned on Muruganantham one sultry afternoon in 1999, when he found his young wife slinking out, 'hiding something'. Curious, the young man wanted to know what she was about. She was evasive at first. Then she admitted how she was disposing off old rags she had used to contain her menstrual flow. 'It was dirtier than the rag I used in the workshop.' Muruganantham realised at once, using rags was a great health hazard for women. In fact many studies conducted by various organisations have revealed that unhygienic conditions during the periods are one of the major causes of cervical cancer amongst rural women.

'I asked her why she did not use the commercial napkins available in the shops. She told me, "if all of us women in this one house use the commercial napkins, we will have to cut the milk budget"', Muruganantham recalls.

Commercial sanitary napkins are costly, a pack of ten costing anywhere between ₹30-40. They are needed every month, year after year, for thirty-five to forty reproductive years in a woman's life. Recent findings under the 'Sanitary Protection: Every Woman's Health Right' reviewed and organised by Plan India reveal that 70 per cent women in India cannot afford sanitary napkins.

In India, at least 400 million women need sanitary napkins every month. According to social scientists, 200 million women in India live Below the Poverty Line (BPL). At least 100 million find spending ₹40-80 a month on sanitary napkins completely unaffordable.

The Indian sanitary napkin market, according to another media study, is dominated by Procter & Gamble (makers of brand Whisper) and Johnson & Johnson (makers of brands Stayfree and Carefree), with Kimberly Clark Lever, a third player. They service only 7 per cent of the population and only 3 per cent of rural Indian women use sanitary napkins.

Yet another recent report says, 'to boost female health and hygiene in rural India' the Indian

government has decided to supply sanitary napkins to 200 million rural women, providing each with a 100-napkin stock for one year and has set aside ₹2,000 crores for this. The recently announced reduction in the excise duty on sanitary napkins from 10 per cent to 1 per cent will also bring about much relief but ten years before the government thought about this, shocked by what his wife told him, Muruganantham began researching sanitary napkins. He first asked himself why was it so costly. It contained cotton. Muruganantham, hailing from the textile district of India, calculated that the 6 gm of cotton inside a napkin should not cost more than 10 Indian paisa.

Then started his low-cost napkin-making experiment, his wife was his first subject. She rejected his first product outright. It was useless, she said. Why was it useless? She did not like his questions on the subject, in rural India, even in the twenty-first century, menstrual cycle is a taboo subject. 'People just don't speak of this.' After a great deal of persuasion, Muruganantham's wife told him, the napkin he had designed was non-absorbent.

He wanted to look at used napkins to study their capacity to absorb blood. He needed more volunteers, he needed actual feedback. When he went around asking relatives to contribute used napkins, his family called him a pervert. His wife threatened to leave him. His mother burst into bitter tears, thinking he had gone mad, when she found him in an outhouse, secretly studying used sanitary towels in great detail. A fancy laboratory in sterile conditions! Innovation in India often has to take a leap of faith over huge social hurdles, as Muruganantham did.

He thought if he asked girl students in the city's medical college to fill in feedback sheets and collected these sheets every two weeks, their responses would give him a clue. 'One day, I found, just two girls hurriedly filling up all the forms, just before I came to collect them. They had been doing this for weeks. So, despite my painstaking effort at collecting data, I had ended up with false data.' Nobody would tell Muruganantham how sanitary napkins were used, how they worked.

'With the help of a friend,' Muruganantham then tells us, 'I filled a football bladder with animal blood, tied it between my legs, and spent days wearing napkins to find out exactly where I was going wrong.'

Finally, after several years of effort, he managed to enlist sixty volunteers for his cause. He gave them his test napkins, commercial napkins, feedback sheets and carry-bags in which they returned him the used napkins for study.

In the process, he also learnt English, began using the internet and searching online for information. It was 2002, almost three years after he began his effort to produce hygienic low-cost sanitary napkins that Muruganantham discovered that commercial napkins contained cellulose, from pine wood, which gave it the absorbency it has. This his napkins lacked.

He told foreign firms in the business that he was a mill owner and wanted to diversify into the sanitary napkin business. Soon samples of cellulose boards, which were not made in India due to unfavourable climactic conditions, arrived at his doorstep. He learnt that a machine that could make the napkin from cotton and cellulose board would cost ₹50 crores. This is when Muruganantham's talent as an innovative machine-maker came in.

In another two years, he had made wife and sisters a machine that would make them clean napkins. He soon found, they were selling single napkins to neighbourhood women. This in Tamil Nadu, India's most urban state with over 70 per cent literacy rate.

Muruganantham then set up his napkin machine company called Jayashree Industries. One machine costs ₹65,000, and it takes three weeks to build. One machine can make 1,000 napkins in a day. 'The napkins are bigger in size, cost just about ₹1 to make and can be customised.' The cellulose has to be imported. With raw material, it takes about ₹1,50,000 to set up a napkin-making unit.

In 2006, Muruganantham approached IIT Madras to evaluate his machine. The IIT innovation team, preparing for its annual festival, send him back a questionnaire in English to answer about his innovation. 'I told them, could they please take answers in Tamil. They were very gracious and translated the questions into Tamil for me. I answered their questions and forgot about it. Several months later, I learnt that my machine was judged the best innovation in a competition that saw 689 entries.'

Muruganantham's innovation has been patented by the NIF. Muruganantham is, however, very clear that his machine would have to help poor, rural women. He, therefore, gives it only to women's self-help groups that produce sanitary napkins for local needs and allow women to make a living out of these machines.

As many as 300 of Muruganantham's machines are today in use in rural Andhra Pradesh, Kerala, Karnataka, Maharashtra, Gujarat, Uttar Pradesh, Bihar, Assam, Nagaland, Jharkhand, Chhattisgarh, Uttarkhand, Sikkim, Delhi, Puducherry, Haryana, Rajasthan and Madhya Pradesh. He also gets orders from Kenya, Ethiopia, Nigeria, Zimbabwe, Israel, Bangladesh, Nepal and Sri Lanka. 'Poor women in

all these countries face similar difficulties in using sanitary napkins,' Muruganantham points out. They are all bound together now by one man's untiring quest for their empowerment and well-being.

—Papri Sri Raman

A Muruganantham in his workshop, working on his low-cost sanitary napkin-making machine. This semi-automatic machine produces 120 sanitary napkins an hour and runs on a single phase power supply.

BlackBerry
MarketWatch
SENSEX
20044.07(137.97)
Symbol
ABB
ACC
IFCI
RIL
WIPRO
TCS
BSE300
6714.99(0.0)
Change %
-0.53
0.15
0.80
0.58
2.43
0.44

Portable Pathways to
Empowerment

Late in joining the PC revolution, millions of Indians are poised to leapfrog into the mobile-and-connected era by embracing the mobile phone and other portable platforms for empowerment and social uplift.

When he visited India for the first time in 2007, Vint Cerf, the 'Father of the Internet', the man who crafted the TCP/IP protocol that is at the bedrock of the internet's 'language', took time off to share with the infotech community here, his own perception of a Web-enabled future. Asked how India could catch up, when its usage of the Personal Computer was so low, Cerf had a terse suggestion: 'Forget the PC', he said, 'it will soon be history. The mobile phone is the personal appliance of the future.'

Three years down the road, his suggestion turns out to be remarkably prescient—Indians have embraced the mobile phone in numbers that have astounded industry and analysts alike—over 16 million new phones are connected every month. By mid-2010, over 800 million mobile phones were in use which meant almost three out of every four Indians had primary—and personal access—to a communication tool.

Access to the internet was still limited to less than 100 million (40 million of them through mobile phones) but the number was galloping because the largely private cellular phone service providers had learnt their lessons on the older second generation (2G) GSM and CDMA systems. They knew that the money they had collectively paid (US$ 22 billion) for getting the licenses of the new third generation (3G) and broadband wireless access, could never be recovered unless, they provided internet on mobile phones as cheaply as they had provided voice calls. The world mocked at Indian telecom providers who worked for Average (monthly) Revenue Per User or ARPU of a dollar or less. They may not do so again, when these same providers roll out mobile-based web-driven value-added services for just a little more.

A study by KPMG analysts in September 2010, suggested that Indian consumers—even those challenged by literacy—were in many respects, more mature takers of technology than those in more developed geographies. Despite global concerns over privacy and security, 38 per cent of Indian customers polled, said they already used their phones to shop and 43 per cent used it for banking transactions.

Not surprisingly, Indians have created some of the most innovative solutions that harness the mobile phone and other portable and hand-held devices to reach the newly connected millions in the rural hinterland with solutions in branchless and micro banking; access to market information and facilitation for E-trading.

The challenge of taking banking services to villages which did not have even one bank, is being overcome by niche developers and voluntary agencies who have whipped together a combo of cutting-edge, yet 'appropriate' technologies.

Zero Banking

A Little World, a Mumbai-based not-for-profit agency has created a Made in India—for India solution called simply 'Zero'. Zero enables new generation mobile phones to work as a secure, self-sufficient bank branch in rural areas which do not have the physical infrastructure of a bank. To ensure easy access amongst the rural population local Customer Service Points (CSP) have been set up where an official issues special smart cards to the villagers on the spot. These smart cards are Radio Frequency Identification (RFID) enabled and use biometrics technology. The customers also get the print out of the transaction, as the mobile phones can be easily attached to the printer, thereby making the entire banking process transparent. Pensioners in Warangal district, Andhra Pradesh and in Mizoram in the North East were among the first in India to be provided with smart cards for smooth transactions.

The Zero Banker has been deployed by the Zero Micro Finance (ZMF) and Savings Support Foundation in the north eastern states, in Uttarakhand and Andhra Pradesh with great success. Apart from the State Bank of India, which has used the solution to fuel payments under the Mahatma Gandhi National Rural Employment Guarantee Act, the Zero solution is today being used by fourteen other banks including Dena Bank, Andhra Pradesh Grameena Vikas Bank, Punjab National Bank, Axis Bank, Punjab & Sind Bank, Oriental Bank of Commerce and United Bank of India.

A Mobile Finance Services Terminal

In 2007 an Indian portable financial applications terminal took the GSM Association's award for Most Innovative Mobile Application in a Vertical Market at the Asia Mobile Congress in Macau. Called iMFAST (Integra's Mobile Financial Applications Secure Terminal), the rugged device that allowed banks to operate in unserviced rural areas by harnessing smart card technology was crafted at the Bengaluru labs of Integra Micro Systems.

The iMFAST system offers all the advantages of the current day mobile banking environment to the villager who is semi-literate and faces problems in commuting to a bank's branch in his vicinity. The terminal's services have also been a boon for the villagers as they no longer have to depend on moneylenders and their savings are safe in the national banks. A commercial bank can link this

device to its banking system using a variety of communication channels like Public Switched Telephone Network (PSTN), or cellular communication and extend all banking facilities to population in rural areas without opening remote branches or installing expensive ATMs.

The iMFAST terminal ensures security as it uses RFID technology for identification and biometrics for authentication. The device also acquires online secure confirmation for the transactions made and thus facilitates safe transactions at the cost of a mobile phone call. The device uses normal phone connectivity for transactions and in areas where there is no phone connectivity available, it can record all the transactions offline and transfer them onto the server at a later time, which makes it more flexible to use in 'unconnected' rural areas of the country. The terminal provides voice guidance support in the local language so that customers can self-operate it with minimal help.

Canara Bank and Corporation Bank have deployed this system. The latter has gone further than many other banks in its branchless banking initiatives particularly in South India. It has used its portable banking system in specific verticals like milk procurement in Tamil Nadu.

Spoken Web

The use of the spoken word as the primary means to reach and harness the internet has been a particular concern of researchers at IBM who have now perfected the technology called the 'Spoken Web'. It has been harnessed with success by the Gujarat Cooperative Milk Marketing Federation, which signed a deal with IBM to use the Spoken Web to let its suppliers access market information on the web in their language, without having to use a keyboard of any kind. In late 2010, Bharti Airtel which won a pan-Africa contract, to launch mobile services in that continent, signed up IBM to provide the Spoken Web among other technologies to help bridge the language gap in half a dozen African nations.

Use ATM without a Card!

The mobile phone is also being seen as a substitute for an ATM card. An Indian start-up, 3P Technologies has created technology called 'BankOn' which works like this:

The ATM displays need to be modified to allow for a non-card mode of operation. If I want to send you (living in another city or for that matter, country) a thousand rupees or its equivalent, I use the BankOn service to send you (on your mobile phone via SMS), a unique four digit code, as well as telling you how much money I am sending. You enter this code at the ATM, and the amount you want to draw (it must

exactly match the amount I am sending you). Through BankOn, I have also indicated the mobile number of my beneficiary and the STD/ISD code of his/her country/city. The transaction goes through only when ATM used matches the city, the amount sought to be drawn is exactly same as the one authorised and the unique code sent only to the recipient's phone tallies. 'It's fool proof and it throws open the ATM network to millions who don't even have a bank account,' claims 3P Technologies' CEO Shourabh Shrivastav. He is working on a voice mode of his solution, that will allow visually challenged bank customers to safely draw money from ATMs the card-less way. Modifying a standard ATM to meet the Reserve Bank's new stipulation that every third new ATM should be disabled friendly typically costs around ₹50,000. The 3P solution suggests this can be achieved more cost-effectively, even while growing the ATM's usage beyond bank customers.

A GSM Tower in Six Hours

For mobile technology solutions discussed earlier, to make a national impact, the challenge of putting rural India on the cellular map remains. The Gurgaon-based Vihaan Networks Limited has created WorldGSM—a cost-effective solution to set up rural mobile networks, speedily and cheaply. WorldGSM mobile tower claims to be the first full-fledged mobile infrastructure that's completely independent of the power grid. The entire system comprising of a base station, batteries, GSM antennae, backhaul and two small-sized solar panels can run on solar power, thereby ensuring environment-friendly use of energy for economic development.

The WorldGSM mobile tower priced at ₹8 lakhs, almost one-fourth the price of traditional GSM tower can also create low-cost Wifi network in the rural areas of the country. Each tower can provide the network over a 500 metre radius and five or six such towers can cover the radius of about 6 kilometres. WorldGSM hence complements existing GSM networks, extending them to embrace the rural opportunity without having to lay cables.

Another positive feature of this tower which makes it rural and environment-friendly is that the entire WorldGSM base station packs into boxes and can be transported by simple means such as a cart or boat. Maintenance is near zero—the solar panels require dusting once or twice a week and the entire set-up comes with a twenty-year guarantee. Software updates are done remotely and any local hardware problems are solved by simple swap repairs.

State-wide SMS Short Code

Sometimes the most effective solutions turn out to be the most simple-minded. Kerala which has the highest telecom density in India has registered a unique SMS short code—537252—across all thirteen telecom districts of the state. This enables the state to offer a wide spectrum of services, from Secondary School Leaving Certificate (SSLC) results and agricultural alerts to emergency requirements of blood and citizen complaints to the police. A bulk SMS gateway has been created which allows government agencies to send instant alerts and communications in many verticals like media, farmers, etc. in a push and pull mode. Other departments that harnessed this M-governance backbone were the Kerala State Road Transport Corporation (KSRTC) for bus route information, Kerala State Electricity Board (KSEB) for electricity billing, Kerala Water Authority (KWA) for water bills and a dedicated website of Kerala Tourism for audio guides to places of interest. In the pipeline were mobile applications for land tax collections, public distribution system and accelerating check post operations.

Some local bodies have gone even further, reaching their citizens with everyday services offered at the end of an SMS or an email. Kozhikode residents can call a phone number; send an SMS or place a request on the website to get a service like plumbing, carpentry, coconut tree climbing, refrigeration and air conditioner repair, TV and electronics, electrical wiring or repairing and gas stove repairing, within one hour. A BSNL landline number; a twenty-four hour automated SMS receiving system and a user-friendly website have been provided. The request is immediately registered and an acknowledgement is sent to the customer's number. The software handles these service requests and assigns appropriate qualified professionals from the available pool. Members from a well trained team of service professionals reach the spot at once. All these services are provided at a nominal cost of ₹100 and a transport fee of ₹50 per house visit.

With 21 million of the state's 31 million population having a mobile phone, nearly 80 per cent reach was assured by any mobile phone-based service in Kerala. By 2011, the state hopes to achieve 100 per cent tele-penetration. But the state may be the harbinger of change on a larger scale—nationwide—as Indians embrace the mobile phone to remain connected, even as they seek to improve their quality of life in every which way that technology will allow.

—Anand Parthasarathy

Mumbai Dabbawallas

Clad in a white kurta pyjama with a Gandhi topi on his head and tiffin boxes in his cart, the *dabbawalla* has become a quintessential icon of the Mumbai culture. The famous *dabba* service of Mumbai which includes timely delivery of home-cooked meals to office goers working in various parts of the city was actually started in 1880. A proper tiffin service was started in 1890 by Mahadeo Havaji with 100 men working with him. The service has grown multiple times since its initial days. Today around 2,00,000 lunches are delivered around the city by more than 5,000 *dabbawalla*s. They charge a nominal fee ranging from ₹250-300 per month making it one of the cheapest options to get hot and healthy food on your table.

Even though they deal with such large numbers, the *dabbawalla*s are known for their punctuality and almost zero error rate. According to a survey, there is only one mistake in every 6,000,000 deliveries which is a record in itself. They follow an in-house coding system that makes the system foolproof. It is this impeccable record that made the *Forbes* magazine declare that the *dabbawalla*s of Mumbai match the six sigma standards. The work force mainly comprises of illiterate or barely educated men and a few women, but they have been given the ISO 9001: 2000 certification. It has been mentioned in the Guinness Book of World Records and is registered with Ripley's Believe It or Not.

However, in place of basking in the glory of its achievements, the association of the *dabbawalla*s has adopted new methods and technology to attract the new age customer. They have launched a website where people can place an order for the pick up and delivery of their tiffins. In this age of cut throat technology the *dabbawalla*s may sound a bit archaic but the facility which they provide covers one of the most basic needs of the people of a very busy metropolitan.

*Dabbawalla*s participating in a computer class at the Dabbawala Abhyas Kendra Study Centre in Mumbai. Most of the men are school dropouts and can only communicate in Hindi or Marathi, but now they have the chance to upgrade their knowledge by picking up basic English and computer skills. Harvard Business School has included this network as a case study, so that its students can learn how a 120-year-old organisation continues to flourish by relying simply on human effort and innovation.

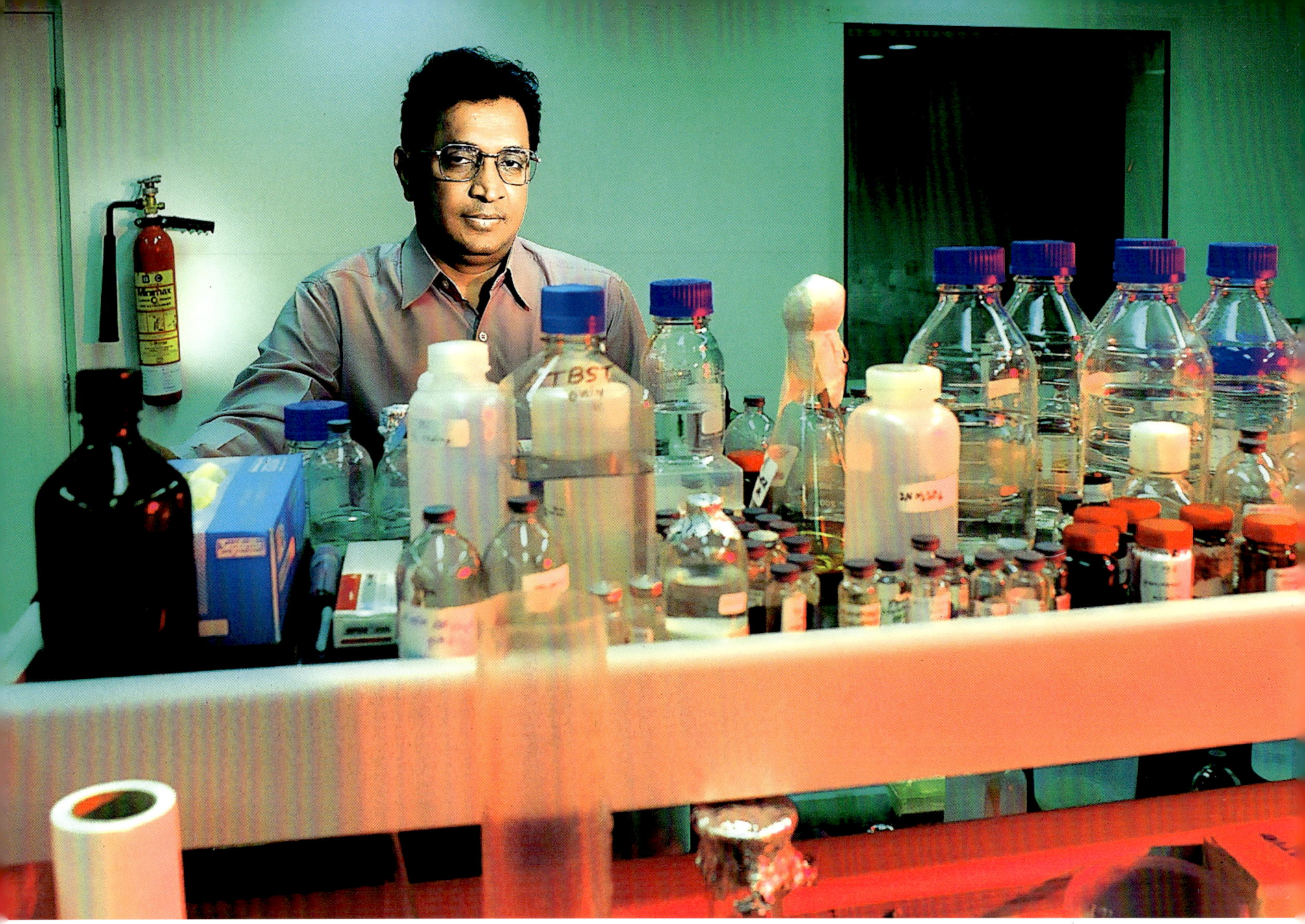

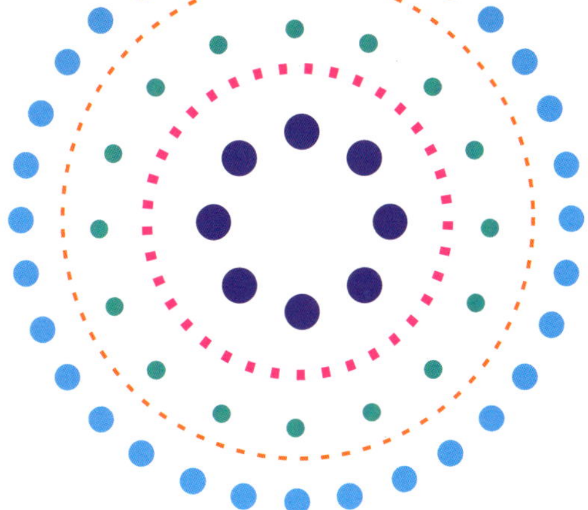

A Stubborn, Die-hard Optimist
Immunising the Poor

Anger, frustration, betrayal, monetary loss, and a hurt pride can be a lethal combination, and has indeed led to many a downfall. Only a few have ever successfully channelised deep anguish to turn adversity into opportunity. KI Varaprasad Reddy is one such person.

The year was 1993. I met Varaprasad Reddy in my office for the first time. He had come to meet Dr PM Bhargava, the well-known biologist who had conceived, built and directed the Centre for Cellular and Molecular Biology (CCMB) in Hyderabad, for advice. Varaprasad, barely concealing his angst at losing both friend and fortune after being unfairly ousted from a flourishing batteries business by a trusted partner, bluntly stated that as an electrical engineer he knew batteries in and out, but knew little about biotechnology beyond its spelling. Yet, he dreamed of, someday, indigenously producing recombinant vaccines in the country. All he had was ₹5 crores to invest. If it wasn't for his steely resolve, one would have scoffed at his sheer impudence.

Cut to 2003. The scene was the runway of TIFR Balloon facility on the outskirts of Hyderabad, where a biology-related experimental balloon was being launched. The retired head of a giant pharma MNC who was also present for the launch exclaimed with total disbelief, 'I cannot imagine I am in India.' He had just arrived there after a whole day at Shantha Biotech. Driving from the glistening, sterile, sprawling 40,000 square feet state-of-the-art vaccine manufacturing facility of Shantha Biotech to the high-tech balloon facility, which several developed countries wait in a queue to use, he was at a loss for words after what he had seen.

Through the decade, Varaprasad had not only done and delivered what he dared to dream of, but also gone ahead at a pace that even he had not prepared himself for. Within five years of promoting Shantha Biotech, in 1997, his privately held company named after Varaprasad's mother, successfully developed India's first genetically engineered product, Shanvac-B vaccine against Hepatitis-B virus. Thus Shantha Biotech became the first company in India to develop, manufacture and market a recombinant product in human health care industry.

Some measure of obstinacy, it seems, is key to turning dreams into reality. Varaprasad's obstinacy was triggered in 1991 at a WHO conference on global impact of immunisation, in Geneva, where he heard mocking references from the West to India's inability to deal with its vaccination needs. Hepatitis-B was turning out to be the biggest killer in India. With only three companies in the world manufacturing the vaccine against the disease, its price was obnoxiously high even for India's middle class, let alone the poor. And India was holding out a begging bowl. Varaprasad returned from the conference with a stubborn resolve that he was going to make vaccines. Six years later when the product was launched, it was the same obstinacy with no rationale that made him resolve to market the vaccine at ₹50 against SmithKline Beecham's product at ₹780.

When Shantha started off on its uncharted path, biotechnology industry was yet to be born in the country. There were neither governmental regulations nor policies. A hilarious incident, Varaprasad recalls, is that when his initial application to the government did not evoke any response, nor did several subsequent reminders, his file was traced in the engineering department; since no one understood the word 'genetically-engineered', it was sent to PWD!

Challenges and obstacles kept him company throughout in myriad ways. His first challenge was to put together a team that could be adequately trained in cutting-edge molecular biology techniques. Acutely aware of his technical inadequacy, Varaprasad relied on friends to fill the gap, and passionately set about being the facilitator.

The next obstacle was to house them in a decently equipped lab in India, but money and man-power were constraints for a stand-alone lab. A brief period of tenancy at Osmania University provided them continuity, and a more impacting and longer tenancy at CCMB, not only came as a rescue but also as a life-saving incubator to his premature baby.

Challenge also came in the form of venture capitalists and bankers. While venture capitalists wanted unrealistic high returns to make up for losses elsewhere, bankers would not mind giving him the money if he pursued proven technologies or doing more of the same. For them funding innovation and that too in an area that had not even taken birth in the country's manufacturing scene, was sheer madness and foolhardiness. An electrical engineer talking of entering the elite world of biotechnology product development and manufacture, and projecting an enormous market size of 325 million children requiring two doses each, was simply wishful thinking.

Respite sometimes comes from the most unexpected quarters. In this case, it was a chance event that led Varaprasad to HE Yusuf Alawi Abdullah, the foreign minister of the Sultanate of Oman, who was looking for a safe place to park his money. Nothing could have been more opportune. Moreover, the money came without strings of interference or management control. The rest of the required funds were pooled in through every other resource he could garner from family, friends and relatives. Like a man obsessed, Varaprasad left no stone unturned and no opportunity untapped to fulfil his mission. In 1997, Technology Development Board, just created by the Government of India to help commercialise home grown technologies, gave Shantha ₹15 crores at 5 per cent interest.

With a low-cost product, another challenge surfaced—no one was willing to market it. An obstinate Varaprasad decided to market it on his own. What ensued soon was a price war with SmithKline Beecham slashing its prices and increasing doctors' incentives, followed by unfair slander about Shanvac-B.

The hallmark of a leader is to find unusual solutions to unusual problems. Varaprasad appealed

to the Indian Medical Association (IMA), Lions Clubs, Rotary clubs, Red Cross and NGOs, offering them the vaccine at an unimaginable cost for mass vaccination. He ear-marked 10 per cent of his production free for orphans through the Ramakrishna Mission. This kind of empathy with the poor and humane approach, overlooking one's own profit, and his invoking national pride through the swadeshi tag, did not go unnoticed. Rich dividends came in the form of Shanvac-B becoming the fastest growing brand. Soon Shanvac-B outperformed competition in every respect— immunogenicity, safety and price. Shantha started supplying a substantial share of the global procurement of UNICEF.

A whole array of products rolled out in quick succession, from a range of combination vaccines (Shan4 and Shan5) to therapeutic proteins such as interferon (Shanferon) and erythropoietin (Shanpoietin). Shan5 is India's first indigenously developed liquid pentavalent vaccine against five major childhood diseases—diphtheria, pertussis, tetanus, Hepatitis-B, infection from hemophilus influenza type B. Recognition for pioneering work, excellence and achievement came in the form of National Technology Award, and the coveted Padma Bhushan from the President of India.

Shantha's crusade continues with their current emphasis on developing ten new vaccines against deadly diseases infecting children, such as rotavirus, cholera, dengue, typhoid, cellular pertusis and human papillomavirus (HPV). Having recently patented a human monoclonal antibody against non-small-cell lung cancer, they have also filed three new patents for other monoclonal antibodies, which have undergone preclinical trials and are moving towards clinical trials.

Shantha's incredible product line tempted Merieux Alliance, a major pharma firm from Europe to acquire a 60 per cent stake in the company. But in 2009 even that would change, with Sanofi Aventis, another global major, acquiring 96 per cent shares of Shantha—a company that started with ₹18 crores for ₹3,007 crores.

Shantha's story is the stuff dreams are made of. Varaprasad has demonstrated both innovation and social entrepreneurism, by developing, producing and marketing health care products at affordable costs without compromising quality. Inspired by Shantha's success, several companies have ventured into this area. Biotech policies have been framed, regulations have been laid, and biotech parks have sprung up in virtually every state—all post Shantha. Thanks to Shantha, at least, no new application with a genetic engineering label on it will go to the table of a clerk in the PWD.

Having gone through and survived difficult labour pangs, Varaprasad's word of advice to financial experts while making appraisals of innovative projects, is to create a column with a heading social benefits and give it due weightage, only then we will create a conducive environment for social entrepreneurship. Commercial viability is the altar on which many such socially spirited entrepreneurs continue to be sacrificed.

Shantha's story explodes two myths: one, that social entrepreneurship and business cannot go hand in hand; two, that lack of a professional label or expertise acts as a barrier to achieving the impossible.

—Chandana Chakrabarti

(facing page) An exhibit depicting Ayurvedic surgery at an exhibition in New Delhi. The famous second century scholar, Sushruta, is considered to be the father of surgery in India. He gave a detailed description of cataract surgery, stone diseases, among others, in his treatise, *Sushrutsamhita* and also mentioned around 121 surgical implements used during an operation.

A scientist working in a lab at Shantha Biotech. In 1997, within five years of its establishment, Shantha Biotech manufactured Shanvac-B, India's first genetically engineered product.

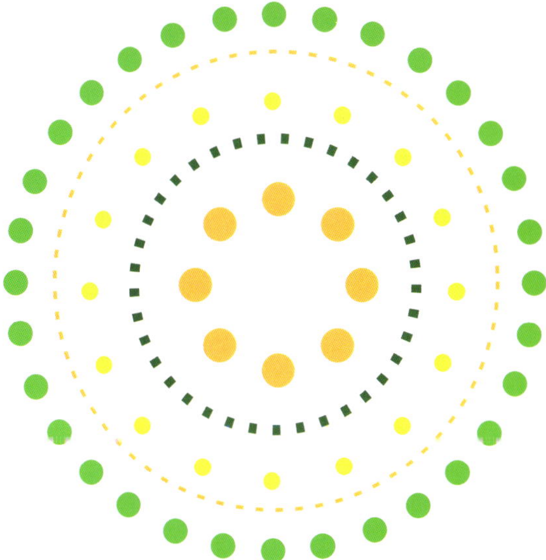

Green Footprints of a Leather Research Lab

The two men shivered in their shoes before an audience of a few hundred—experts, hospital staff, patients and their families—who had gathered to watch a man wearing just one shoe, take it off. The shoe was on the still-surviving foot of a man who had lost the other foot to diabetes.

'We were unprepared for the results,' admits BN Das, who heads the Shoe Design & Development Centre at the Central Leather Research Institute (CLRI) in Chennai. Six months before, the centre had been asked by Chennai-based MV Hospital for Diabetes to design a shoe for a patient. Under the guidance of Ali Foster, a consultant podiatrist from King's College Hospital, London, the CLRI scientists had designed one shoe and 'had forgotten about it'. Six months later, when Das and fellow scientist Gautham Gopalakrishna were sent back to witness their handiwork, 'We did not know what to expect,' says Gopalakrishna. 'We are not doctors.' However, that day in the mid-nineties changed the lives of the scientists of the shoe design centre. The wound in the single surviving foot of the patient had healed considerably.

'It is not a magic shoe,' warns Das, relating the scientists' experience at the Delhi Trade Fair pavilion 2010, where CLRI had put up a demo stall. 'People, some from as far away as Meghalaya, just thronged our stall, asking for the footwear. People think, just the footwear will bring about a magical cure.' It is not that. 'However, protection for a vulnerable foot helps in saving it,' Gopalakrishna points out. The scientists cite the case of a medical professional, who used to recommend amputation for an infected foot, before coming across CLRI's specially designed footwear. Wearing that shoe makes a marked difference in the wound.

'Before they helped the diabetes patients, the shoe design centre scientists had interacted with the Schieffelin Institute of Health Research & Leprosy Centre, Karigiri, to design shoes for leprosy patients,' Gopalakrishna says, 'but we took this as social service.' CLRI now has a memorandum of understanding with the MV Hospital, a premier diabetes care centre in India, with an intellectual property right for seven years to provide design and technical support for footwear prescribed by this hospital. Of India's estimated 51 million diabetics, at least 25 per cent are affected in the foot.

(facing page) An Indian mother poses with her daughter in contrasting outfits. The mother wears traditional Indian garments, while the daughter wears a black, sparkling leather body suit.

The CLRI design centre has contract research agreement with the Novonordisk Education Foundation, a trust set up by a Bengaluru-based pharma company, to provide technical support for developing shoes for ankle-foot orthosis (AFO) in India. Imported footwear may cost up to ₹30,000 but the CLRI shoe costs between ₹1,500 to ₹2,000. The Bhagwan Mahaveer Viklang Sahayata Samiti (BMVSS), Jaipur, the organisation which provides the Jaipur Foot, has also been consulting the design centre for standardisation of their models.

Another design project of the CLRI responded to the needs of the artisans making Kolhapuri chappals and made many poor cobblers exporters of fancy footwear. Kolhapuri chappal is handcrafted ethnic footwear manufactured by rural artisans. In Athani village in Karnataka, CLRI's design centre demonstrated a techno-social innovation through quality and product standardisation, new designs, training and women-led social empowerment. The project introduced standardisation and size consistency. The technology and design inputs could be used because of UNDP support and Bengaluru-based Asian Centre for Entrepreneurial Initiatives (ASCENT)'s endeavours. The NGO organised production and international marketing and established the brand name 'ToeHold' for products made by rural artisans. This handcrafted footwear began to be sought by the fashion conscious customers in many foreign countries. New footwear designs led to a dramatic improvement in the living standards of 400 artisan families of Athani village making the Kolhapuri chappal. On an average a family makes about a 1,000-2,000 pairs in a year. Local cooperatives and women self-help groups make 3,000 or more pairs annually. ToeHold collects these and exports them to buyers all over the world. The village has been transformed. The villagers have access to clean drinking water, medicare and the internet for business purposes. The children go to school and college.

CLRI's innovations and technology delivery systems have transformed India's traditional leather industry, making it environmentally friendly and globally competitive. These have been inclusive innovations. T Ramasami, Secretary, Department of Science & Technology, Government of India says it is important to recognise the process and purpose of innovations that must be made accessible and affordable to the masses. Mahatma Gandhi's vision of an indigenous leather industry guided Ramasami's earlier work as the Director of CLRI. He recalls Gandhi's words written in 1934 in *Harijan* that the export of hides and skins from India and the import

of articles made from these means not only a material but also an intellectual drain. Indians miss the training in tanning and manufacturing articles of leather needed for daily use. 'Here is the work for cent percent swadeshi lover and scope for harnessing of technical skill for the solution of a great problem.' Gandhi lamented that 'tanning has consigned millions of people to hereditary untouchabilty' and he saw in leather related technologies a tool for narrowing social inequities.

The basket of technologies developed and delivered for the rural sector includes better carcass recovery and by-product utilisation; improvements in rural tanning practices and value additions to rural leather products, says a former CLRI scientist. A Amudeswari, Director of the Indo-French Centre for the Promotion of Advanced Research, studied how the S&T innovations resulted in community development. CLRI's technology packages are designed to suit low investments and applications in a rural environment, she says. For example, inputs of technology in the footwear industry created some 50,000 jobs for women with limited education in Vellore district, Tamil Nadu in seven years.

The leather industry in India is spread over Tamil Nadu, Kanpur, Kolkata and Jalandhar. Tamil Nadu manufactures upper leather for footwear, while Kanpur produces sole, Jalandhar manufactures sports leather goods and Kolkata manufactures leather accessories. It provides employment to around 3 million people; of this 30 per cent are women.

Like textile and the beverages industry leather is also a water-intensive industry. As 70 per cent of the leather before tanning is treated with low-cost, toxic chemicals hence the pre-tanning operations lead to pollution. Processing mainly is done in Tamil Nadu (60 per cent). Thus, places like Ranipet (near Vellore) gained notoriety as the most polluted place in India. In the eighties, the government found that leather technologies used in India were not people and environment-friendly and would soon be rejected by the world. The groundwater

(facing page) Embroided leatherware with traditional patterns from Jaisalmer, Rajasthan. Rajasthan's leather products earn considerable foreign exchange. The city of Bikaner in the state is famous worldwide for its consumer goods made out of camel hides.

Leather being processed at a leather factory in India. CLRI, in its bid to adopt eco-friendly techniques for leather production in the country, is leading initiatives to blend natural fibre-based materials from pineapple leaf, banana and soya, with leather products.

table had gone down considerably wherever there was leather industry, every kilo of processed leather needed 50 litres of water. Huge amounts of salts too were required for preservation. Hence, groundwater, rivers and surrounding soils had become extremely contaminated by chemicals near leather processing hubs. 'To save livelihoods, a research organisation like CLRI had to think out of

the box,' says scientist J Raghava Rao of the CLRI Chemical Laboratory.

The Supreme Court ordered the closure of 400 tanneries in Tamil Nadu in 1996 which would have affected 2,50,000 jobs. Sound environmental technologies were adopted and the jobs were saved. Today, the leather sector in Tamil Nadu is able to use Zero Liquid Discharge technologies and yet remains cost competitive in the global market. 'Europe and the USA—India's leather markets, began to show preference for enzymatic unhairing of the skin, green chemicals, eco-labels, pastel shades and eco-benign, bio-processed products. Therefore, it was essential for the industry in India to use high-value, high-performance chemicals in post-tanning and salt-free preservation. All this meant a paradigm shift in processes and CLRI provided the leadership in sustainable technology. We had to pioneer the wealth from waste models and research for zero emission,' Rao points out. It is this turnaround that has resulted in US$ 3,400 million in export earning for Indian leather. In the recent past, scientists and researchers at CLRI have developed a novel technology for leather processing. This 'green' technology uses enzymes to reduce the discharge of harmful substances by 90 per cent during the tanning and pre-tanning steps of leather processing.

CLRI's green expertise has made an impact on the leather industry of many foreign countries. Common Effluent Treatment Plants (CETP) have been set up in Bangladesh and Pakistan. China's leather industry has benefitted by recruiting experts from India. A large number of Indian scientists have joined the leather industry abroad and CLRI has trained more than 2,000 foreign scientists. An Indian laboratory's global green footprint keeps expanding!

—Papri Sri Raman

Cricket balls being stitched at a factory in Jalandhar, Punjab. Jalandhar is one of the main sports manufacturing centres of northern India, and cricket balls made here are exported to Sri Lanka, Australia, Bangladesh and Canada, among other countries.

(facing page top) The tradition of hand-stitched water bags known as *mashaqs* is very old in the subcontinent. One can trace it to the early days when humans learnt to store water in leather bags. Even today, this craft of making leather water bags is alive in some parts of India, with water-bearers, *bhishti*s carrying on the tradition of providing water to the thirsty.

(facing page bottom) Protecting the feet of the world. Today, footwear manufacturing technology in India is based on cutting-edge, computer-aided design facility, bringing Indian producers at par with the global level.

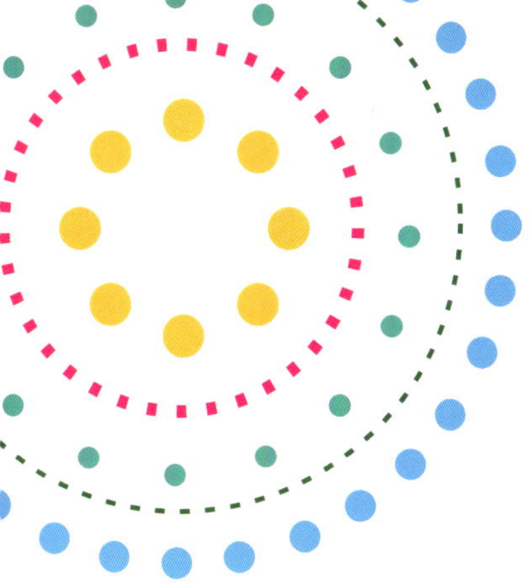

Devi Shetty—
Uniting the World with Cardiac Care

The location: Eritrea in Africa, near the River Nile bordering Chad and Sudan. A young man and a doctor are seated in a small consulting room in a government public health centre. The clock indicates that the time is 7:30 am. The India clock shows 12:30 pm. In Bengaluru, India, cardiac surgeon Dr Devi Shetty sits in his blue surgeons' clothes looking at a monitor placed in front of his table that is beaming the duo from Africa into his hospital room. The two sets of people across continents are now connected through a satellite link and ready to begin a unique medical consultation powered by technology.

The young patient is introduced to Dr Shetty by the doctor in Eritrea. He is twenty-seven-years-old and his name is Maricos. He has been complaining of shortness of breath, a chronic cough and general swelling of the body. He has already undergone the necessary cardiac investigations in his country and he has been diagnosed as a case of 'valvular heart disease'. This consultation is to get advice on the future course of treatment for him.

Online from India, Dr Shetty takes a look at his echo-cardiogram report. He reconfirms the diagnosis and says that one of the valves in Maricos' heart is not functioning well and this needs correction. Over the next ten minutes a discussion takes place between both teams on whether a surgery or replacing the defective valve will help the patient as well as the risks involved.

'Since Maricos is young, the chances of his benefitting from the surgery are high—an early surgery is advisable to avoid chronic heart failure,' says the Bengaluru surgeon. The patient who is an IT employee in Eritrea is advised to travel to Narayana Hrudayalaya for his surgery. The team seeks a few more clarifications regarding the case and the impending surgery and then both sides sign off. The telemedicine link between Bengaluru and Eritrea is successfully completed.

Next it is time to go down to a live link in Bangladesh. Clad in a white kurta Noorjaman, seventy-two, is seated in a consulting chamber with the centre's coordinator, Naheeda Alam. In 1993 the patient has already had an angioplasty done, Dr Shetty is informed.

'*Aapnar oshubidha ki hochche?*' (What is your complaint?) asks the cardiac surgeon in Bengali. Years of working in Kolkata has made the doctor from Bengaluru proficient in Bengali. The elderly patient reports that he gets a pain in his chest when he walks.

Online, once again, Dr Shetty takes a look at his angiography report and the latest echo report. And then comes the verdict in Bengali, '*Ami aapnar report dekechi, aaapnaar heart khoob bhalo ache…khali ekta choto block ache…kono chinta nai…aapni eksho bochon obdhi bachben*' (I am seeing your heart reports, your heart is in a very good condition…you have no reason to worry…you will live till you are 100!).

The elderly Noorjaman breaks into a broad smile almost as if he has been given a new lease of life by the doctor's pronouncement. The patient's wife, who has so far been anxiously peering over her husband's shoulder and listening to the entire consultation on satellite, looks even more relieved. Noorjaman is told to do a follow-up after one year and advised an oil free, red meat free diet. The link from Bangladesh is now done.

Crossing borders, crossing continents, technology is firing health care in an innovative manner. Telemedicine with Africa, Bangladesh, and around 100 district hospitals in India through links provided by ISRO and also a Ministry of External Affairs sponsored US$130 million link through Tele Communication India Limited (TCIL), has made the Bengaluru-based Narayana Hrudayalaya one of the prime hubs for this innovation. The hospital is now fully committed to reaching out over land, sea and air to bring cardiac care to patients in far-off places.

The Pan-African E-network project link was possible because of a vision statement of former Indian President Dr APJ Abdul Kalam. In 2004, at the inaugural session of the Pan-African Parliament held in Johannesburg, Dr Kalam had proposed a programme to connect India with all the fifty-three countries of the African Union (AU) through satellite and fibre optic network.

The philosophy behind it was that India could share its expertise in the field of health care and education with Africa. On the ground, it meant getting a submarine link up from Mumbai all the way to Morocco. In the first phase in May 2008, thirty-one countries in Africa got the telemedicine link. Presently the number has climbed to fifty-three. Apart from treating patients, it also allowed the opportunity to do Continuing Medical Education (CME) with doctors and paramedics in

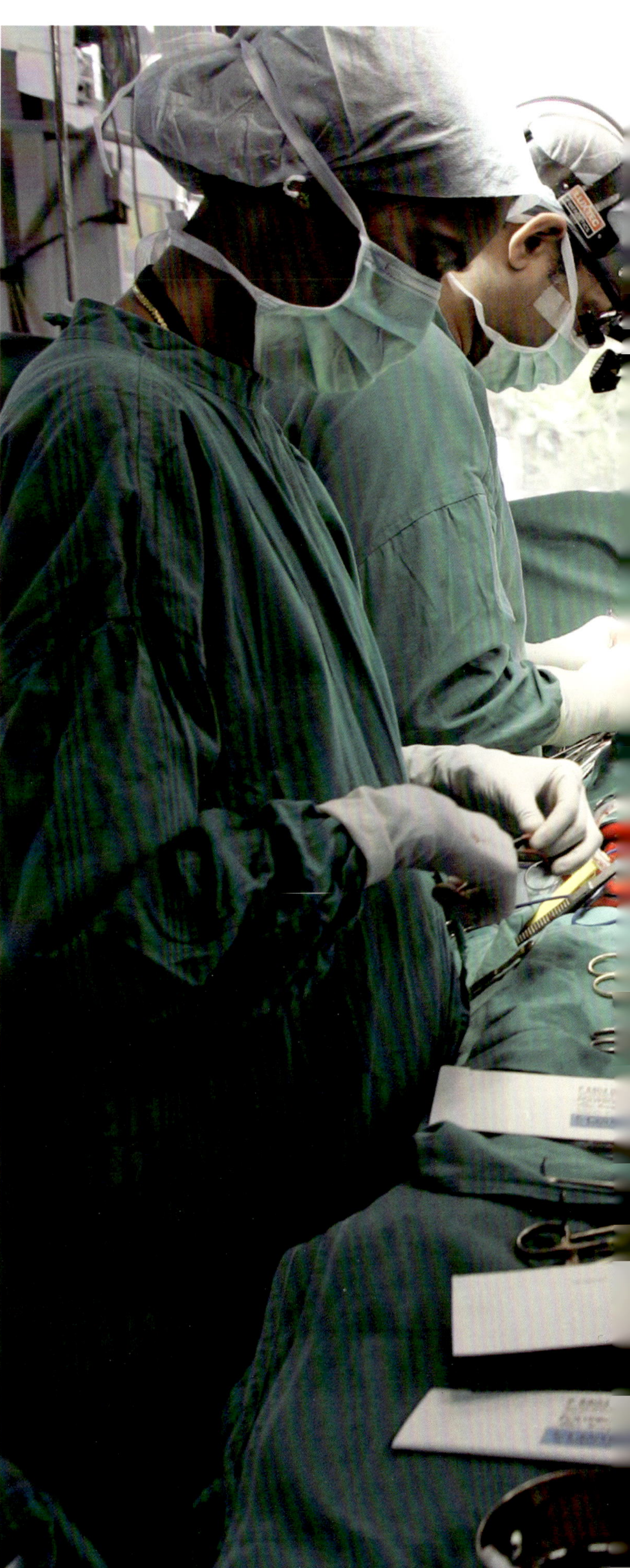

Open heart surgery is carried out on a baby at the Narayana Hrudayalaya. The institute has the largest pediatric cardiac surgical intensive therapy unit in the world to look after patients, especially children, who have undergone heart surgeries. Cardiac surgeons at the hospital are renowned for carrying out operations on newborn babies, particularly related to the complex heart problem known as 'transportation of great arteries'.

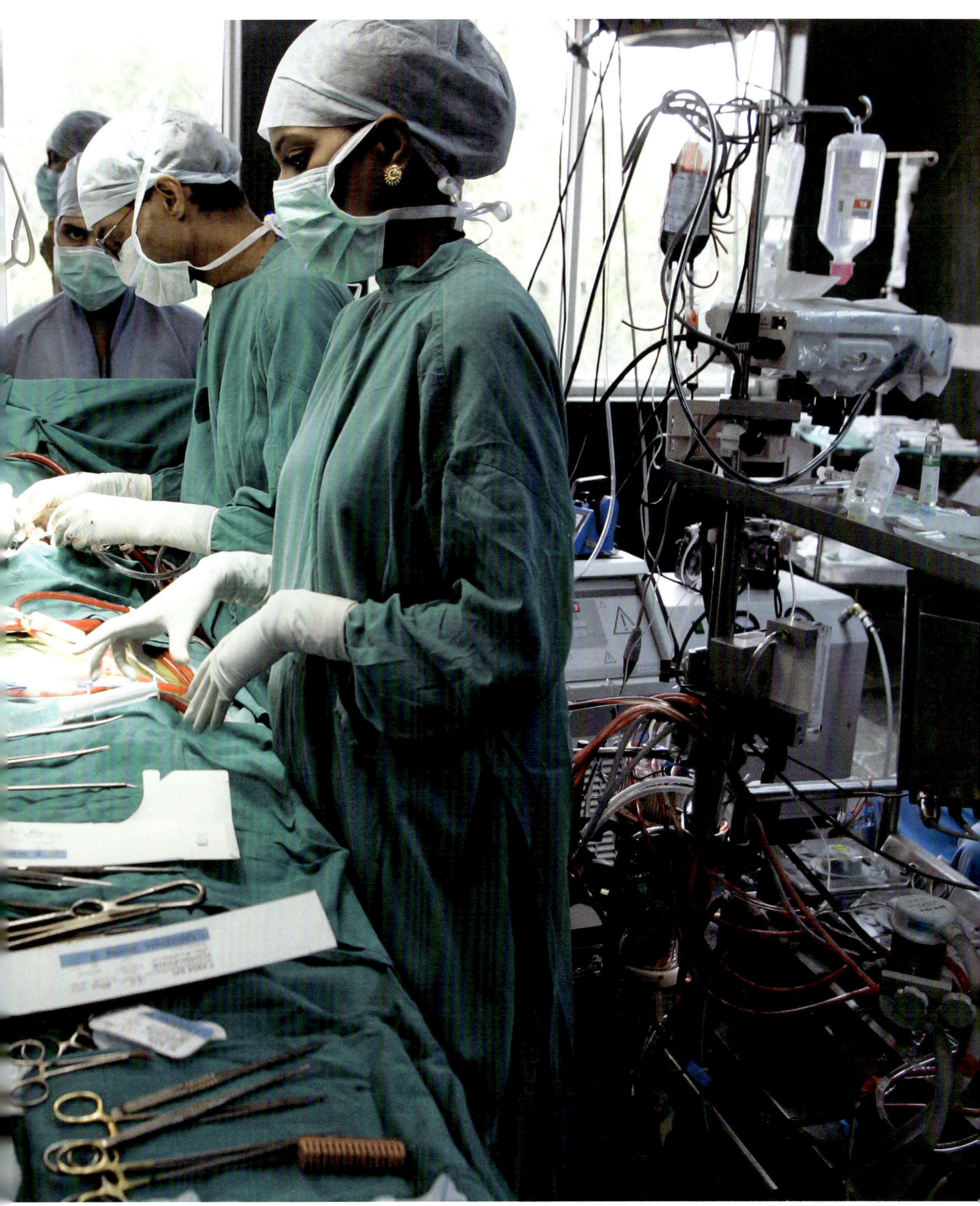

Seychelles, Botswana, Nigeria, Ghana, Gabon, Gambia, Mauritius and other African countries. 'If India has to become a superpower we are not going to be able to do it with weapons. If we have to be treated as a superpower, we have to be able to hold the hand of other less developed nations and make them walk,' believes Dr Shetty whose role models are Mahatma Gandhi and Mother Teresa.

Telemedicine is a familiar tool for this doctor who saw the potential of this technology eight years ago. When Dr Shetty moved from Kolkata to Bengaluru, his patients clamouring for him in Bengal, soon found a good way to stay connected. ISRO gave the transponders and Dr Shetty could now be in touch with his patients from Kolkata through satellite technology. 'Satellite links are distance insensitive and for us it was a necessity to embrace this technology,' he recalls.

Having had the privilege of being Mother Teresa's cardiac surgeon in Kolkata, he narrates a story that had a great impact on him. 'One day Mother Teresa was walking in the children's ward in the cardiac unit with me and she said, "when God created these children he was probably preoccupied—so he sent you to fix it later," I can never forget what she said.'

The complicated heart surgery of Noor, a small girl from Pakistan by Dr Shetty's team at Narayana Hrudayalaya had brought into national focus, for the first time, the heart surgeon's efforts to breach borders. People loved the story and the media pulled out all the stops to cover it. This, in turn, resulted in an entire nation rallying and praying for the young girl from Lahore, even as the surgeons operated. When the operation gave a new lease of life to little Noor, it opened the floodgates for 'get well soon' cards and cuddly soft toys posted to our television news rooms for Noor from ordinary people in India. A little girl's fight for survival had united two warring nations.

Subsequently hundreds of mothers, fathers, daughters and sons flocked from Pakistan to the Bengaluru hospital to seek the treatment for their loved ones, which they were unable to get in their own country.

Dr Shetty and his team scored on diplomacy more than any of India's politicians or diplomats with Operation Noor. Although this had also opened the doors for telemedicine with Pakistan, in recent months the poor relations between the two countries has put a stop to the links.

Besides having personally imbibed compassion from Mother Teresa, Dr Shetty also brings to the table a business acumen of a true blooded Mangalorean, a region in India known for their banking and educational enterprises. His shrewd business brain and management innovations have made his cardiac care delivery model a success story in terms of economies of scale.

In comparison with the cost of an open heart surgery in the US, which costs anywhere between US$20,000 to US$100,000, his 1000-bedded medical facility charges approximately US$2,000, thereby giving cardiac care treatment in India a comparative advantage. Known to drive hard bargains for surgical equipment and hospital supplies with providers with innovative purchase policies, Dr Shetty believes in the proverbial robbing from Peter to give to Paul.

By bringing down his hospital overhead costs, he ensures treatment at a price far less than even the other cardiac care units in the private sector in India. He told *Wall Street Journal* in an interview once, 'Japanese companies reinvented the process of making cars and that's what we are doing in health care—what health care needs is process innovation, not product innovation.'

Presently Dr Shetty's dream project is to build ten low-cost heart hospitals across the country so that the poor in India can have better cardiac care. 'India is a poor country with a few rich people and the greatest breakthrough in health care are low-cost heart hospitals,' he says. Hospitals which would cost ₹80-100 crores are being done within ₹13-14 crores. According to him only 8 per cent of the world's population can afford heart surgeries. 'In India, doctors are able to perform only 80,000 to 90,000 surgeries every year although it is estimated that 2.5 million people require heart surgeries,' Shetty points out.

Personally this cardiac surgeon is committed to chase the Africa dream in the long-term. 'After sixty I am going to move to Africa and dedicate myself to bringing cardiac care there.'

For inspiration to achieve his plans riding the latest innovation, he dips into Mother Teresa's advice, 'Give the world the best you have and it may never be enough. Give the world the best you've got...anyway.'

—Nupur Basu

(facing page) Dr Devi Shetty is considered to be the Henry Ford of heart surgery. He has proved how a new age model in health care, with state-of-the-art connectivity, can be set up on sound economics—get more for less. Narayana Hrudayalaya has established video and internet links with 500 telemedicine centres spread across the world, giving expert advice and saving lives.

Mobile ECGs

A large population of Indians lives in villages where health centres and doctors are not easily accessible or well equipped. For those with heart morbidity in rural India who need an instant ECG in crises, there is now technology available to get this report. Narayana Hrudayalaya started an ECG network whereby hundreds of ECG reports are transmitted through telephone lines. With 600 million mobile telephone users in the country, mobile ECGs have become a new buzzword in medical technology innovation. Efforts to get wired up are beginning to show results.

Dr Devi Shetty emphasises, 'Indians are genetically three times more vulnerable than Europeans. Indians develop heart attack at a much younger age. A heart attack must be diagnosed within six hours and treated in a period called "golden period of heart attack". Generally, when the heart attack strikes, most villagers do not get an opportunity to get the ECG done. In the process they sustain irreversible damage to the heart and a lot of them die. Our intention is to diagnose heart attacks even in remote locations of India so we tie up with the rural general practitioners and give them a trans-telephonic ECG machine made by Schiller from Germany. Costing about ₹10,000, this device can record the ECG and send it by telephone to our hospital in Bengaluru. Everyday we get hundreds and hundreds of ECGs and our doctors report ECG within ten minutes of its arrival. ECG which comes to our telemedicine room is linked to our coronary care unit and it is seen on any of the computers on the corridor of the coronary care unit. Any doctor walking by can report them. Everyday we diagnose large number of heart attack incidences and the general practitioners give appropriate advice to the patient. We would have saved at least few thousands lives by this simple technology.'

Yeshasvini

This landmark initiative by the government of Karnataka and Dr Devi Shetty was inaugurated on November 14 2002. It is the world's largest scheme of self-funded health care and provides health care to the poor farmers of the state. All its members have access to expensive treatments and medical procedures covering more than 1,600 surgical procedures. The beneficiaries of the scheme are also offered cashless treatment at more than 135 hospitals of the state.

The scheme has attracted attention from a number of international institutes including the Harvard and the Rockefeller Foundation that are studying this scheme closely to replicate it elsewhere, especially in the African countries. The International Labour Organisation has also highlighted Yeshasvini on its website and has conducted a study as well. The World Bank has also used the mechanism of this scheme to find more practical solutions to low-cost but high quality health facility in the developing countries.

The Hills are Alive
to the **Sound of Water**

The urban tourists, fascinated by the scenic beauty of the hills, rarely notice the problems of the isolated communities living there. The villagers cope with hardship and disasters, natural and man-made—deforestation, dried up streams and rudimentary health and transport facilities. They and their crops have little protection from adverse weather conditions. They do not have the help of the young who migrate to towns in the plains.

In the hilly areas of Uttarakhand, two innovative projects have brought much relief to the villagers. One has revived dead and dying water springs. The other has modernised watermills, increasing their efficiency and the earnings of those running these. In addition to grinding grains, these mills are now used to generate electricity. As if some one exhorted the villagers to 'Keep the Streams Flowing' and 'Keep the Wheels Turning'.

It all began in 1981 when a young college lecturer wanting to work for rural development noticed that the locals facing day-to-day problems of existence had neither time nor energy to look for long-term solutions. Nor were they too favourably disposed towards a visiting outsider-benefactor. He understood that environmental sciences, simple technologies, traditional wisdom and local resources could help resolve some of the problems of the hilly areas.

Dr Anil Joshi, who was born in a small town at the foothills of the Shivalik range of the Himalayas, established the Himalayan Environmental Studies and Conservation Organization (HESCO) based on the Gandhian principle of self-reliance at the village level. He decided to live in the midst of the villagers to win their trust. His team that includes four PhD holders works from village Shuklapur near Dehradun. The 'Mountain Man', as he is popularly known, has demonstrated how the combined use of scientific knowledge, traditional wisdom, simple technology and local skills and resources can transform lives. HESCO's contribution has impacted more than 5,00,000 villagers in 10,000 villages spread over eight hilly states. It has won him national and international recognition including the prestigious Padma Shri, the fourth highest civilian award given by the Government of India.

(facing page) Tour boats waiting on the shores of Bhimtal Lake, Bhimtal, Uttarakhand. Innovative projects have revived dead springs in the region, proving to be a boon for villagers as these water bodies are a primary source of sustenance for them.

Silent Springs

When a water spring in the hills of Uttarakhand, which sustained the villagers for ages, dried up, some villagers perhaps prayed but Dr Joshi called for the help of BARC. He approached a scientist who had recently read a paper at a seminar on isotope hydrology. It is a nuclear technique that uses both stable and radioactive environmental isotopes to trace the movements of water in the hydrological cycle. Thus began a new partnership and a model project with encouragement from Dr R Chidambaram, former Chairman of the Atomic Energy Commission. The shared objective was to revive the dead spring. Tracing its recharge area was a critical task. The scientists employed, for the first time in India, the environmental isotope technique to look for the sources of springs in order to construct artificial recharge structures for rainwater harvesting and groundwater augmentation.

The people of hill states depend heavily on the springs for water. In Uttarakhand, some 10,000 of the 16,000 villages have been affected by water scarcity. Deforestation has drastically reduced the catchment areas while poor rainfall and regular earthquakes have further worsened the situation. Of all the springs in the state, 30 per cent have almost dried up and 45 per cent are on the verge of drying up. This has affected more than half of the rural population of the state. The villagers have to walk long distances to meet their daily needs. Sometimes, clashes also take place over water disputes.

A spring dries up mainly because the rain water that charges the catchment starts running off due to deforestation. If the rain water can be harvested by creating a conservation facility in the catchment area, it will percolate slowly to the natural aquifer and nourish the stream again. The environmental isotope technique helps to trace aquifers in the catchment area that used to feed the spring. Once these are traced, recharge structures are built, connection made and the dead spring returns to life.

BARC scientists were able to demonstrate this through a model project in the mountainous region of Gaucher area in the Chamoli district of Uttarakhand. The recharging project began with the recording of the discharge of springs, sending of the spring water samples to BARC laboratory in Mumbai. Rain gauge stations were also set up in the hills to collect the rainfall data, and the samples were sent to BARC. The vegetation was surveyed. The isotopic composition of precipitation is affected by season, latitude, altitude, amount and the

distance from the coast (continental effect). In order to find the sources of the selected springs, upper-catchment zones were surveyed, and rainfall samples from these were collected for isotope studies. The isotopes measured in the water samples from these were then matched to the isotopes in the water found in the springs. The areas where the isotopes matched with the spring water were identified as the probable recharging zones for the springs.

Apart from isotope information, the scientists factored in the data related to the local geology, geomorphology and hydrochemistry. On these selected sites the subsurface dykes, check bunds and the contour trenches were built for controlling subsurface flow for rainwater harvesting and groundwater augme ntation respectively. Creeping grass was planted in the catchment area so that it absorbed the torrential rainfall flow and ensured slow and steady recharging. This water percolated and recharged the natural aquifers that fed the springs.

The model project was successful. During and after the following monsoon, the discharge rates of the springs increased significantly and did not dry up even during the next dry season. The project demonstrated the use of isotope techniques to identify recharge areas of the springs in the Himalayan region. These techniques provide information that cannot be obtained by other means.

The miraculous increase in water discharge has astonished the community! As a result of the Gwar catchment area project, the availability of water in the springs serving three villages went up significantly. Even with a minimum rainfall (600-1200 mm), these springs, with increased discharge, will be able to meet the needs of the people. The impact is visible. Many young people have started commercial cultivation of vegetables in areas where vegetables were never cultivated before. Vikram Singh, who was unemployed, earns a decent living today by cultivating vegetables for the local market. Jaspal has begun to use the surplus spring water for a fishery.

The villagers want to ensure that water scarcity does not recur. They have built new tanks in their catchment area for water storage. Spring recharging has thus become a community effort. One villager, Narendra Mishra, says it is a water revolution of sorts.

BARC will set up a facility for isotope studies in the mountains. Its Isotope Applications Division has initiated isotope hydrological investigations in nine

more locations. These will help identify the groundwater sanctuaries and the recharge areas to create artificial recharge by rainwater harvesting. These sites are located in Himachal Pradesh, Uttarakhand, Jammu & Kashmir, Mizoram and Maharashtra. The multiplier effect of the innovation, spreading across regions, touches environment and lives.

Watermills

In Europe, an old watermill and its scenic backdrop reminds the locals of their pastoral past. But in India's hilly villages, a running old watermill, used for grinding grain, provides livelihood to a family. It is the same in many other developing countries. Depending on his mill, Shukal Chand Sangatya barely managed to make a living. Then, he heard of HESCO running a watermill upgradation project. He enrolled himself. Once his mill was technically modified, his monthly earning went up from ₹1,200 to ₹7,500.

Donkwala, an old village situated 8 kilometres from Dehradun, had no access to electricity and other amenities like roads, hospitals and schools. It had three watermills that were used for grinding grain and thus providing a source of livelihood to the owners. If somewhat modernised, these mills could perform additional functions. HESCO offered help to tap this potential. The three mills were modified and made more efficient and the village was blessed with a new source of power—3 kilowatts of it. HESCO went on to implement more projects and in association with the Northern Command of the Indian Army, electrified some border villages. Many more projects are needed to cover some 2,00,000 old watermills in the country serving the rural communities.

The earliest watermills were established far back in the third century. Because of its simple and cost-effective mechanism, the watermill is still popular. Moreover, the villagers had no other alternative even though the watermill is not a very efficient and productive machine. How does it work? A watermill works on the same principle on which a big hydro-electric project runs. The mill is turned by a little stream the likes of which are found in the valleys of the hills. Water from the stream is tapped and routed through a chute. A jet of water strikes the blades of a wheel. The force rotates the wheel which moves the grinding stone.

While upgrading the watermill, HESCO introduced the power generation feature by using metallic turbines which were fabricated locally. The shaft of the turbine is attached to a pulley that runs different agro-processing machines. The ultimate pulley is attached to a generator to get electricity. The turbine is made in two parts so that it can be

easily carried over long distances in the hills. Dr Joshi estimates that if all the traditional watermills are modernised, the hills could produce up to 2,500 MW of power.

A belt arrangement makes a watermill suitable for additional operations with increased efficiency in milling of flour, dehusking, cotton combing, oil expelling, weaving, juice extraction and pulping, etc. In some villages, an improved watermill can be seen running a wheat mill, a saw mill and a loom. The villagers are glad to participate in the construction and maintenance of such watermills. They are impressed by the upgraded mills performing multiple tasks. Clean energy through these mills cuts carbon emission. And since the use of wood is reduced, many trees are saved in the process.

In Himachal Pradesh and Jammu & Kashmir, the watermill is used for grinding maize, wheat and sugar. In the North East, particularly in Manipur, it is also used for dehusking, and is called husker.

In Uttarakhand, the watermill is called *gharat* or *ghat* and is mainly used for grinding grain. The watermill is environment-friendly, a product of local efforts and a part of the tradition as well. It will always have a role as no other energy option can compete with it in many rural areas.

Minor technology inputs in this simple time-tested device show dramatic results. Significantly, these inputs can be provided locally. Following a watermill movement launched in 1990, HESCO called upon the Government of India to help revive the mills on which the hill communities depended heavily. The Ministry of New and Renewable Energy decided to support the millers. Apart from Uttarakhand, these projects have been implemented in Arunachal Pradesh, Himachal Pradesh and Jammu & Kashmir. The upgraded mills constitute a decentralised community service system.

HESCO promoted a society of local youth trained to fabricate watermill parts and agricultural tools. This group also supplies smokeless stoves, solar driers and solar geysers to the rural communities. It has installed more than 160 watermills in Jammu & Kashmir, eighteen in Himachal Pradesh, sixty in Uttarakhand, eight in North East, one each in Gujarat and Jharkhand, doing business worth ₹1,48,80,000.

The project planning took into account the skills, economic status and maintenance capacity of the community. An innovation based on local resources was seen to be affordable and adopted widely. That is why it has been a success. Dr Joshi, who got the help of a hi-tech institution like BARC for his rural development project of reviving dead springs, has a piece of advice for the scientific community, 'It should keep in touch with traditional knowledge.'

—LK Sharma

The Solar Sisters—
Bringing Light to Lives

As night falls, the inhabitants of 111 remote villages scattered in thirty-two African, Latin American, Arab and Asian countries, think of Tilonia in Rajasthan! Their homes are lit by solar power through photovoltaic panels installed and maintained by the mothers and grandmothers of each village who returned from India after a six-month training. These women had left their home for the first time to travel to Rajasthan to learn from unschooled local women to assemble sophisticated equipment. In the absence of a shared language, the trainers taught the learners through gestures and by colour-coding the parts to be assembled—the same way they themselves had learnt earlier from their seniors. Those who come to the Barefoot College in Tilonia, without ever having used a screw driver, go on to fabricate solar panels and install lighting systems. The growing tribe of these trained women settled in villages across the world is called Solar Sisters and Sunshine Warriors.

Mochozi is a grandmother from Congo who does not think it is too late to learn new tricks. Also from Congo is Moyoni, a thirty-eight-year-old mother of eight children. They are among twenty-four trainees of the new batch that includes only one man (from Jordan). Moyoni studied only up to Class IV but is able to speak a few words in English. She likes her stay in the Rajasthan village which is so different from her own. The 'new food is no problem' though she does miss cassava. This batch, like the previous ones, is being trained by the local women and men. From a nearby village of Baori comes forty-five-years-old Magan Kanwar, who got trained in solar equipment assembly and installation in 2004 and then began training others. As a married woman, she once could not go out to work and had started taking up tailoring work at home. That was before she got recruited in the solar training programme. Magan Kanwar says she had never imagined she would get to meet women from so many far-off countries. At times she gets a telephone call from a former trainee from a remote village in a foreign country. Two other local persons, Ganpat and Firoz Khan, are also part of the training programme run by Bhagwat Nandan who had joined the college soon after it was established.

So far some 142 Barefoot women solar engineers have returned to thirty-two countries to light electric bulbs in more than 10,000 houses of 111 villages. The community of a chosen village selects the women candidates whose travel and training costs are paid by the Technical Economic Cooperation Mission of the Ministry of External Affairs, a partnership applauded by the World Photovoltaic Community. The mothers and grandmothers, who return to their countries after training in India and set up the solar lighting system in their respective villages, become empowered, self-dependent and productive members of the community. In 2005, the Barefoot College trained three Afghan grandmothers who went on to maintain Afghanistan's first solar electrified village and to train twenty-seven local women to help keep the lights on. More Afghan women have come to the college since then and the aid in this case also included the supply of solar power equipment to their country.

The Solar Sisters have transformed the lives of the villagers. Batch after batch of poor women, who may not read or write, have been trained in Tilonia as solar and water engineers, midwives, designers, communicators, architects and rural social entrepreneurs. As a result of electrification, boys and girls who graze cattle during the day can study after sun down. Women can earn additional income by doing needlework at night. Perishable medicines can be refrigerated in village clinics. Then there are environmental benefits. A rural family in Africa burns around 60 litres of kerosene a year to light its home. The average kerosene lamp spews a tonne of carbon dioxide in less than ten years. Solar power, by replacing kerosene and wood, improves the health of the environment as well as of the people. Also, it is less expensive. The village women no longer have to walk long distances to fetch kerosene, wood, candles and torch batteries for lighting at high costs.

Bunker Roy, Founder of the Barefoot College, has given thought to every detail of the development strategy. Every traditional society recognises the grandmother's role as nanny, cook and storyteller. But few imagine her handling coils, charge controllers and inverters and connecting solar panels to bulbs via the batteries. After installation, the system needs maintenance and repair and the very same grandmother is there to keep it going. Roy discriminates in favour of grandmothers while recruiting the trainees for the simple reason that

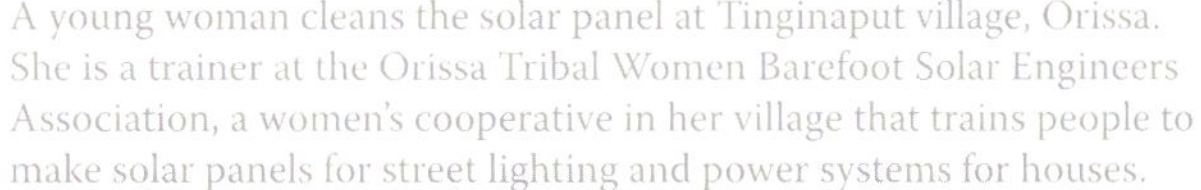

A young woman cleans the solar panel at Tinginaput village, Orissa. She is a trainer at the Orissa Tribal Women Barefoot Solar Engineers Association, a women's cooperative in her village that trains people to make solar panels for street lighting and power systems for houses.

grandmothers are receptive and easy to teach and they have a vested interest in their village which they do not wish to leave. This ensures that the village continues to have a working solar lighting system. Give a youth a piece of paper and he is off to the city to find a better job. Roy's mission is to enable the young men and women to stay in their own villages and improve its environment rather than migrate to cities and end up in slums. The strategy of engendering development goes a long way.

The solar lighting systems are sustained through community ownership and participation and by nurturing local skills. Villagers keep contributing a nominal monthly fee (based on the family's expenditure on kerosene) for the maintenance of the system. They see the benefit of collaboration. Having made its choice in a collective and transparent fashion, the community remains committed to the project and takes all decisions. This removes the major cause of failure of many rural development projects where solutions are imposed from the top. This bottom-up strategy promotes collaboration, empowerment and self-reliance.

A single dollar of the Indian aid programme goes much farther than the one that comes from a Western aid agency. Moreover, the aid penetrates the areas where the international aid effort has difficulty reaching. Of course, India runs large overseas aid projects but this modest rural electrification training programme, covering the least developed countries, is among the most innovative. Bunker Roy has more than three

decades of experience of living and working among villagers. The Barefoot College is the only fully solar electrified college established in a village in India. The solar components (inverters, charge controllers, battery boxes, stands) were all fabricated in the college itself and installed by Barefoot women solar engineers under the overall supervision of a priest educated up to secondary school.

The programme that turns out women solar engineers has shattered many myths about poverty, development, centralised aid programmes and technology. Bunker Roy runs the institution on some of Mahatma Gandhi's principles, recognising the importance of self-reliance and the skills and potential of deprived and marginalised people. The training programme marries modern technology with local skills and demystifies both learning and technology. The learning-by-doing is based on the fundamental belief that the very poor have the right to control, manage and own the sophisticated technologies to improve their own lives. According to Roy, the Barefoot College puts technology in the hands of the villagers themselves. It has demonstrated the creativity of the people who, if empowered, can identify and meet their needs far better than others. Roy says the college has rejected the outside professionals from the formal education system. It believes instead in identifying and using the skills, knowledge and practical experience available among ordinary people within the community to provide for basic needs such as drinking water, health, education, employment, fuel and fodder.

The solar electrification programme has made an impact within India too. In sixteen states, some 383 Barefoot solar engineers including 169 semi-literate rural women have electrified 648 villages, helping 15,000 families and 483 night schools. They have distributed 5,220 solar lanterns. All this has saved nearly 2 million litres of kerosene per year from being used for lighting. These engineers have also installed solar cookers and solar water heaters in the Himalayan region. Near the largest inland salt lake at Sambhar in Rajsthan, the college has installed the first-ever Indian solar-powered reverse osmosis desalination plant to produce 600 litres of potable (sweet) water per hour from the salty water. More such plants are being established.

As the message from Tilonia continues to spread, India will find many more countries wanting to import its solar revolution!

—LK Sharma

Husk Power System

There are a number of path-breaking innovations taking place all over the world today. Scientists and innovators are creating things beyond human imagination but what is it that really makes a difference—a gadget or innovation that reflects the spark of the creator or a simple solution that makes a difference in the lives of thousands of people across geographies.

Husk Power Systems (HPS) is doing exactly that, it is making a difference where it matters the most. In a world where a quarter of the population has no access to electricity they use rice husk, a waste product, to bring light to the lives of many. In India alone, more than 400 million people have no electricity. Once the sun sets, everything is covered in darkness for them. HPS based in Bihar, has created a way to turn rice husk into electricity. This unique method is reliable, eco-friendly and inexpensive. One family needs to spend only ₹80 every month for this facility which serves a total of 30,000 households across Bihar.

The company was founded by four friends: Gyanesh Pandey, Manoj Sinha, Ratnesh Yadav and Charles W Ransler, who met attending different schools in India and the USA. Pandey, the company's Chief Executive, grew up in a village in Bihar without electricity but went on to study at the Rensselaer Polytechnic Institute, New York before landing a job with International Rectifier in Los Angeles. He was living the great American dream when he realised that he needs to return home and use his knowledge to bring light to Bihar.

Once back in India, he tied up with his friend Ratnesh Yadav and started experimenting with the idea of producing organic solar cells. It was during that time that he came across gasifiers which are essentially machines that burn organic materials to produce biogas. These machines were very common in rural areas but nobody even thought of using

them to run a power system. Based on this idea, the two friends started work on an electric distribution system powered by rice husks which was abundant in Bihar. Finally, they came up with a system that could burn 50 kilograms of rice husk per hour and produce 32 kilowatts of power, sufficient for about 500 village households. This system went live on August 15 2007, the anniversary of India's independence.

At the same time, their friends in the USA, Sinha and Ransler, put together a business plan and set out to raise money. They collected money by winning various competitions and also managed to receive a grant from the Shell Foundation to set up three more systems in 2008. Since then they have raised US$1.75 million in investment financing. By 2009, they had nineteen systems in operation and in 2010 they had more than the triple of what they started out with. By the end of 2011, the company expects to have 200 systems, each serving a village or a group of small villages. They plan to set up 2,014 units serving millions of clients by the end of 2014. And to avoid electricity theft or inadvertent overuse, the company is also developing a pre-paid card reader for home use. The card which usually costs between US$50-US$90 will be available for only under US$7.

HPS did not stop at creating electricity out of waste, it went a step further and found ways to use the rice husk *char* which is the residue of a waste product. They use this *char* to make incense sticks, a business which operates in five locations and provides income to 500 women.

HPS has made it possible to deliver electricity to far-flung places while earning a 30 per cent profit margin. The side businesses add another 20 per cent to the bottom line. Each unit now becomes profitable within two to three months of installation and the company is expecting to be financially self-sustaining by June, 2011.

Apart from the financial advantage, the business model has a number of social benefits as well. The shopkeepers can stay open till late, farmers can irrigate their fields any time they want, children can study at night and most importantly, the company employs locals thus helping them earn regular wages. The company is dedicated in providing a long-term solution to the people. They recruit locals with little or no education and train them. They also tied up with Havells India to purchase thousands of high quality bulbs at discount rates, which their collectors now sell directly to clients. These collectors also supply other products such as soaps, biscuits and oil at discount rates.

Countries across the world work on the downward flow of power—from large electricity plants to smaller villages. HPS proved them wrong by creating electricity within a village at minimal cost and by using a waste product found in abundance. Ramapati Kumar, an advisor on Climate and Energy for Greenpeace India, who has studied HPS, says that the company's model could, 'Go a long way in bringing light to 1,25,000 unelectrified villages in India,' while reducing, 'the country's dependence on fossil fuels'.

HPS is a young company and to achieve its goals it will need to recruit and train a number of employees and raise a substantial amount of finances. Whether it succeeds or not, only time will tell but the group of friends has shown that innovation is not always about making new gadgets and machinery, sometimes it is about the simple things. About taking a simple product and creating something extraordinary out of it. The company illustrates a different way to think about innovation—an approach which is suitable for global problems that stem from poor people's lack of access to one of their basic necessities—energy.

AVRA Labs—
the Art, Science and Business of Synthesis

At seventy-five, Dr AV Rama Rao, Founder and CMD of AVRA Labs, walks upright in long strides, talks in a booming voice, and appears just as indefatigable as he did in 1985 when he took on the reigns of a rusty and deteriorating Regional Research Laboratory (RRL) in Hyderabad as its Director. The long, poorly-lit corridors of RRL then were abuzz with whispers about the incumbent Director, whose reputation of being a no-nonsense, straightforward, hard taskmaster was already churning the lab that had for long become used to an easy life.

True to his reputation, in the ten years of his Directorship, 1985-95, Rama Rao turned around a sick laboratory to become one of the best national laboratories by way of its contributions in organic chemistry and chemical technology, in the areas of agrochemicals, drugs and fine chemicals. It was a decade-long journey during which Rama Rao did everything that needed to be done, from the most unpleasant to the most difficult, for a complete metamorphosis of the institution, namely tackling unions; throwing out the bad apples—unheard of in government institutions; instilling a work culture where there was none; and literally changing the name and the face of the institution. RRL got a new name, Indian Institute of Chemical Technology (IICT), a new buzz with hordes of students in the laboratories, a new ethos, a new respect, a new pride, and many new facilities. The transformation was there for everyone to see, in terms of revenue flowing from industries, young scientist awards, Bhatnagar Awards, technology awards, increased publications, patents and industry collaborations.

When the inevitable age of retirement came knocking at his door, Rama Rao, still young, energetic, dynamic, and vibrant, was ready with packed bags to start off on another voyage on an uncharted course. A doting wife, three bright grown-up children, and loving grandchildren would have been considered ideal retirement conditions by any ordinary person, but not so for Rama Rao. He was only willing to retire from government service, but was loathe to retiring from professional life. Right through his career, he looked forward to change every five years. Retirement for him was yet another change, not a stop as it would be for

(facing page) Students getting first-hand training in a biotechnology laboratory in New Delhi. In 2010-11, biotechnology was one of the fastest growing knowledge-based sectors in the country and earned a revenue of ₹15,000 crores. Experts believe that by 2015, this figure will reach ₹45,000 crores, with a 30 per cent growth every year.

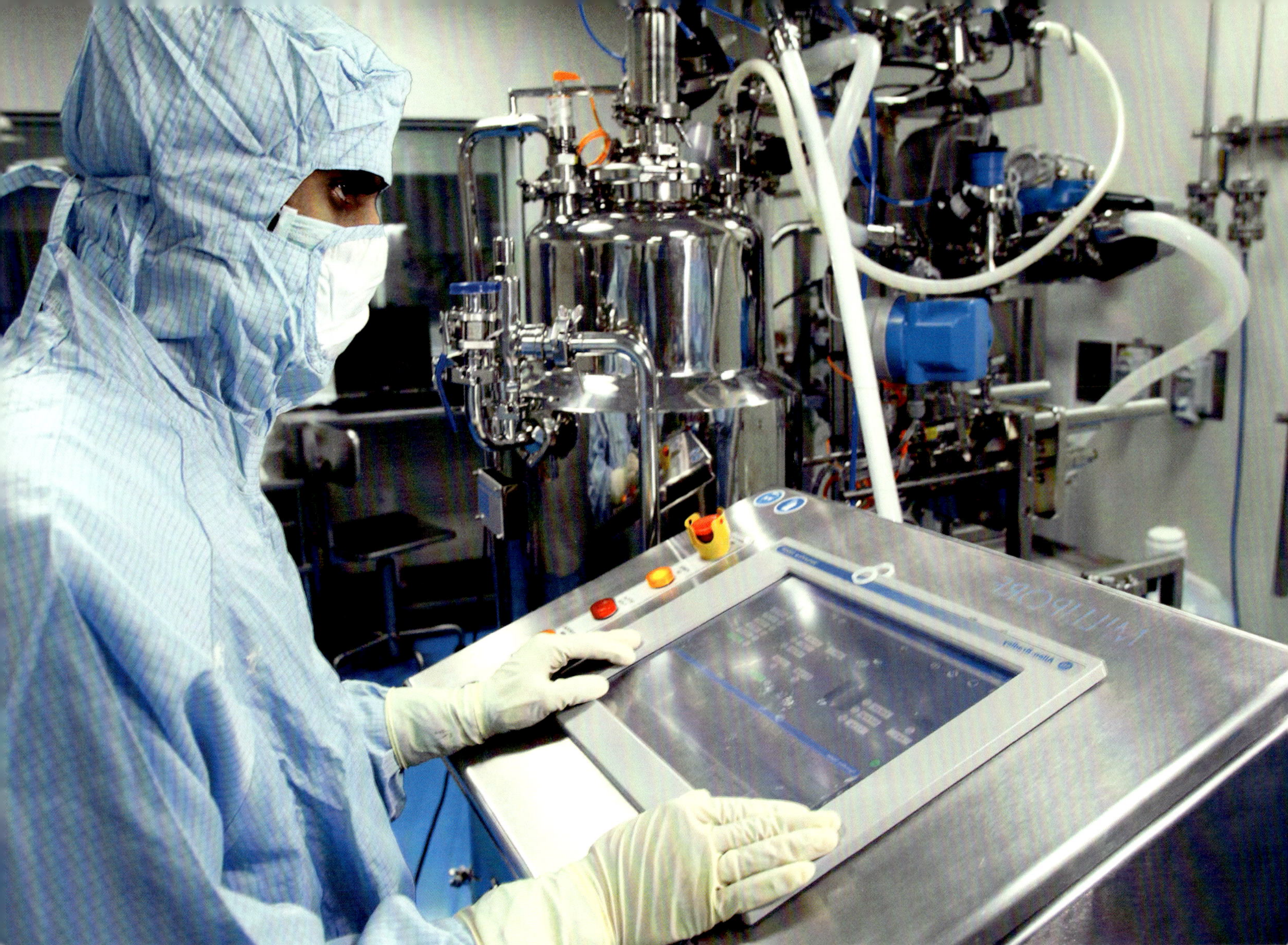

many, 'When I retired as Director of IICT in 1995, I decided to utilise my scientific expertise to tackle some industrial projects. I wanted to offer more than paper solutions; I wanted to have a lab where I could work with a team of scientists and provide real solutions on a fee for service model. This idea led to the genesis of AVRA Laboratories.'

Opportunity came in the form of an offer of an unused godown from DM Neterwala, owner of Dai-Ichi Karkaria Ltd and a challenge from GD Searle, a major drug firm of USA, to develop an alternate process to bypass the nasty smell of a by-product in the synthesis of a drug. Around the same time, while lecturing at the prestigious Gordon Research Conference in USA, a chance meeting with the CEO of Cytomed, a biotech firm, brought him another major but tricky contract to synthesise a new chemical entity, CMI392, that was eluding the manufacturers and others who were given a

similar contract. Such was his reputation of being a clever organic chemist that both the companies paid the money up-front. While the offer of the godown came in handy for a make-shift laboratory, the two contracts saw him firmly in business, and their successful completion brought in a trail of MNC contracts from the likes of Pfizer, Bristol-Myers and Astra-Zeneca. Thus began his post-retirement chapter of a re-invented entrepreneur. He would soon willy-nilly become an icon of outsourced research.

Starting from a contract research and manufacturing services company, AVRA is today involved in early stage research of drug development, synthesising building blocks for new chemical entities, innovative process chemistry, identifying naturally occurring molecules with medicinal properties and then developing a process to synthesise them. An example would be SN-38. AVRA, for the first time synthesised SN-38, an intermediary that can be used to produce irinotecan, an effective and well-known anti-cancer drug. So far this drug had been derived from the naturally

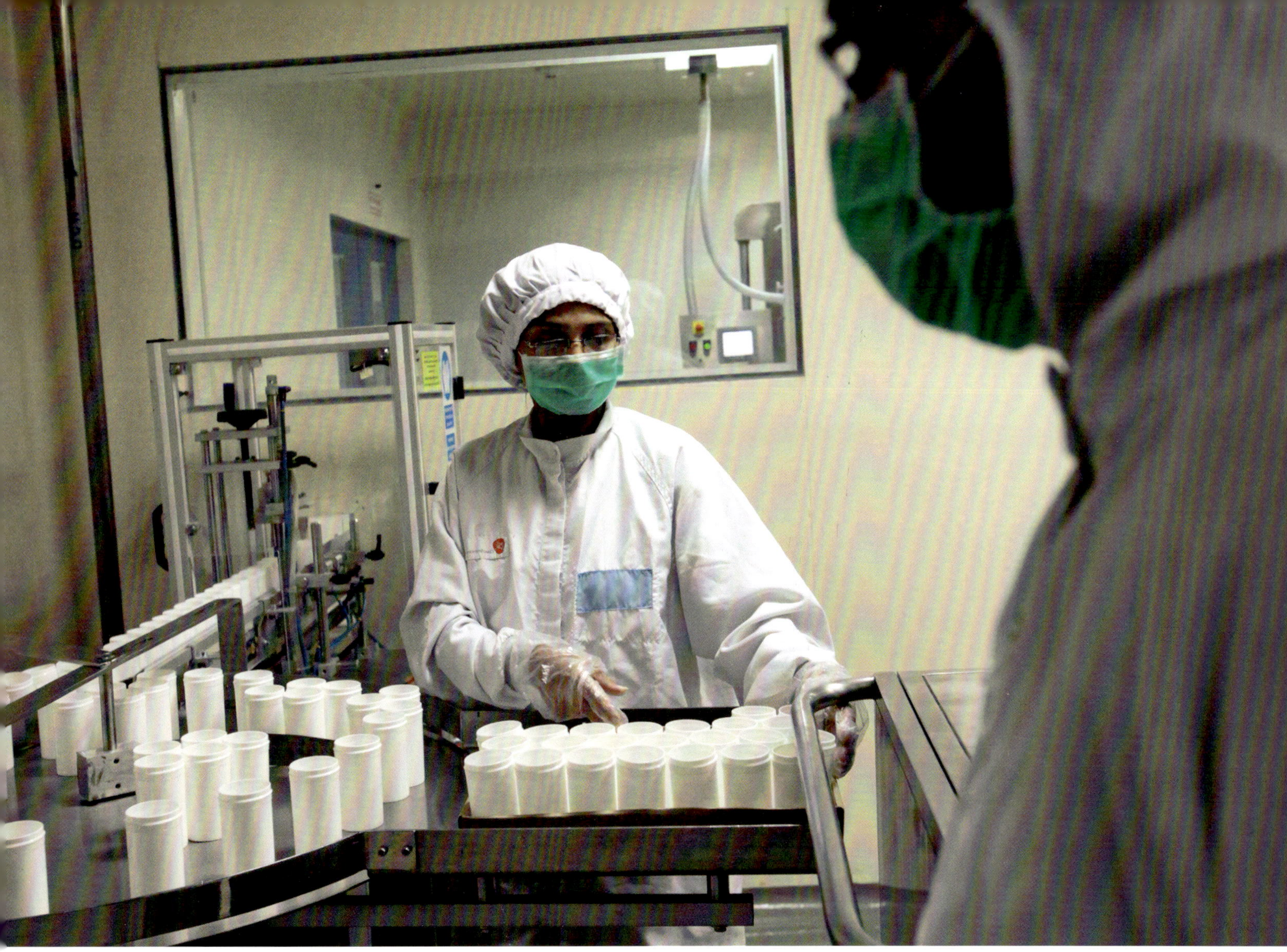

Employees working in the bottle-filling section of a pharmaceuticals company in Nashik, Maharashtra. In 2011, the Drugs Controller General of India (DGCI) decided to introduce online services for the country's pharmaceutical companies—from submission of their applications to getting their licences. This step has been taken to create transparency in drug regulation in the country.

occurring camptothecin extracted from campto-theca, popularly known as Happy Tree and found in China. It is a far more expensive route to making the drug when compared to making it by conversion of SN-38. Further, the synthetic route provides 99.8 per cent pure irinotecan, as against 98 per cent purity by the natural process.

Sitting in his plush room in AVRA Labs now, amidst verdant surroundings that house over 70,000 sq feet of built space that includes state-of-the-art infrastructure for developing and manufacturing in both limited scale and large-scale, Rama Rao says with palpable relief and pride, 'We do not owe a penny to anyone in debt.' Indeed, an unusual, true claim.

This would not be surprising for anyone who even cursorily tracked Rama Rao's innovations all through his career, whether in National Chemical Laboratory (NCL) in Pune or in IICT in Hyderabad. Courage, confidence, hard work, risk-taking ability, never wanting to wait for favourable conditions, getting a high from facing challenges, looking for solutions in the most improbable places, refusing to get bogged down by bureaucracy, inspiring his co-workers, were all his forte. His personal history is replete with such instances.

Rama Rao always attempted to synthesise the end molecule by unconventional means. In 1970, while in NCL he helped a young entrepreneur who started Poona Synthetics, to convert piles of wasted PTS-amide which is a by-product while manufacturing a key ingredient of saccharine, into urethane derivative, an intermediate for an anti-diabetic drug. They sold 100 kilograms of the compound to Hoechst India Limited, which was a tall order given the stringent conditions it had to meet. Besides, he also found its use in making fluorescent pigments that were used for printing textiles and in the production of printing inks. These fluorescent pigments were then being imported from Japan. In all this endeavour Rama Rao not only dirtied his hands in the lab but also

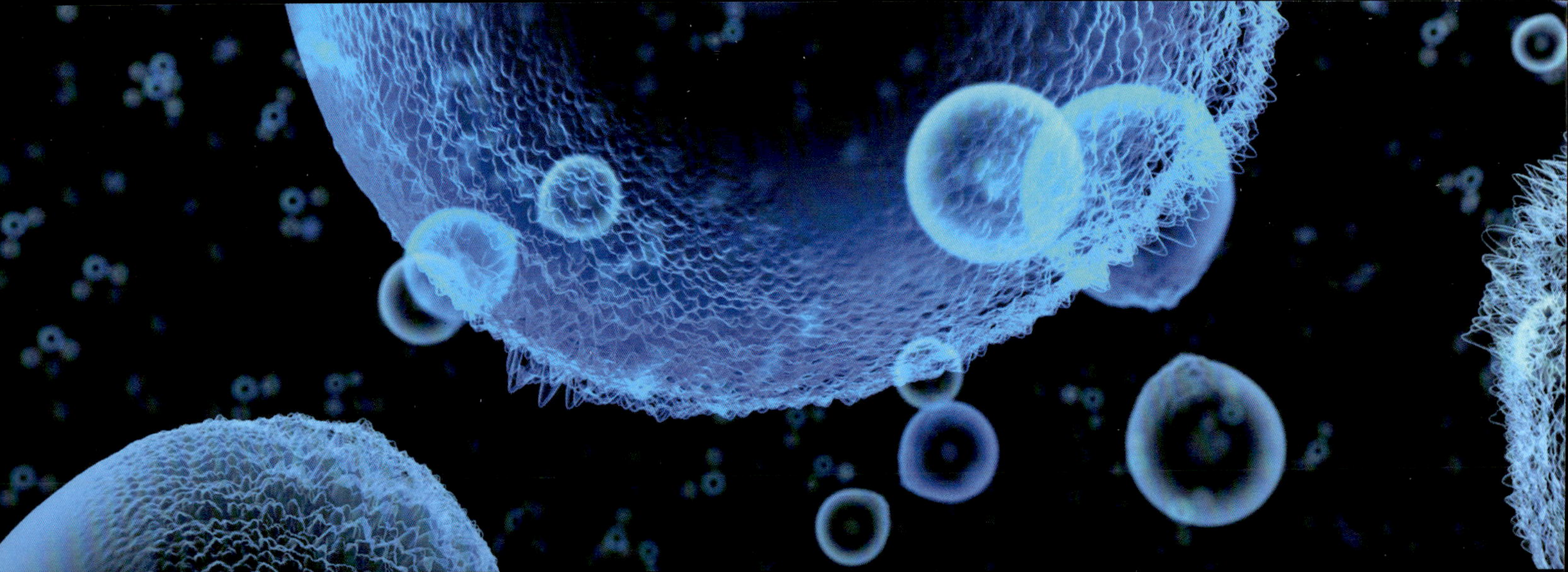

designed the plant and the machinery required, and prodded the bank with assurance of financial viability to get Poona Synthetics the extra finance required. Poona Synthetics thus came out of its troubled times. This was a far cry from the stereotype of Indian scientists working in ivory towers.

In 1972, again in NCL, he developed a process to synthesise the tranquiliser, diazepam, and sold the technology to Centaur Chemicals. Around that time was his first meeting with Dr Yusuf Hamied of Cipla, a major pharmaceutical firm in India, which would lead not only to a technology transfer straightaway but also a productive relationship in the years to come.

A two-year stint at Harvard in the laboratory of Nobel laureate EJ Corey, acquainted Rama Rao with anti-tumour drugs which fascinated him so much that after returning to India in 1977, he decided to work in that area against all odds. Defying conventional wisdom of senior scientists around, elderly advice from the then Director who was clearly unwilling to support him if he went against his advice, Rama Rao set out on the task of isolating vincristine and vinblastine, two anti-cancer agents, from the leaves of *Vinca rosea*, popularly called periwinkle, on a project of Maharashtra government. Vincristine till date is the only effective drug against leukaemia in children. It was not an easy task given that the plant has ninety-five alkaloids, which are generally toxic nitrogen-containing organic compounds found in the plant world. Chromatography equipment was not available then in India, and the yield was painfully small even in the West. Rama Rao, in his inimitable style, found a simple method of isolating vinblastine from *Vinca rosea* leaves, completely bypassing the chromatography method. In place of chromatography equipment, he used solvent and large drums in which he welded taps. Then he went about finding a simple way of converting vinblastine to vincristine by oxidation with potassium permanganate. Not only was it a simpler, quicker and cheaper method, the yield was also higher, a fact that would reflect in its pricing. Rama Rao's persuasive powers made an initially reluctant Cipla buy the technology in 1983. While Cipla sold 1 milligram of vincristine vials at ₹25, an imported vial of the same which was exempted from import duty, was costing ₹86. Not only did Cipla sell it cheaper, it also donated the first 500 vials of the drug to leading cancer hospitals in the country.

Rama Rao's sheer persistence led to the development of a more efficient and cheaper process of synthesising Vitamin B6, which is considered one of the finest examples of innovation. The process was sold to Lupin, another drug manufacturing company in India. He also worked for Cipla, on ibuprofen, a non-steroidal anti-inflammatory drug and beta-blockers such as metoprolol and atenolol given to heart patients, besides salbutamol, an anti-asthma drug. Another area in which Rama Rao made a great contribution to the country and the developing world was in HIV/AIDS. His process of synthesising zidovudine, commonly called AZT, commercialised by Cipla in 1993, brought down the price of the drug to one sixth of the international price then. But that did not come easy. He had to take on the might of the multinational, Burroughs Wellcome and fight our own bureaucracy. Courage and conscience never left his side. When push came to shove, in order to move an official sitting on his file for clearance of AZT, Rama Rao even picked up the phone threatening him to take the matter to the Prime Minister of India who was also the President of CSIR. The file moved and permission was granted. It was solely due to his initial efforts that it became possible for Dr Yusuf Hamied to later offer to the third world countries, the more effective cocktail of

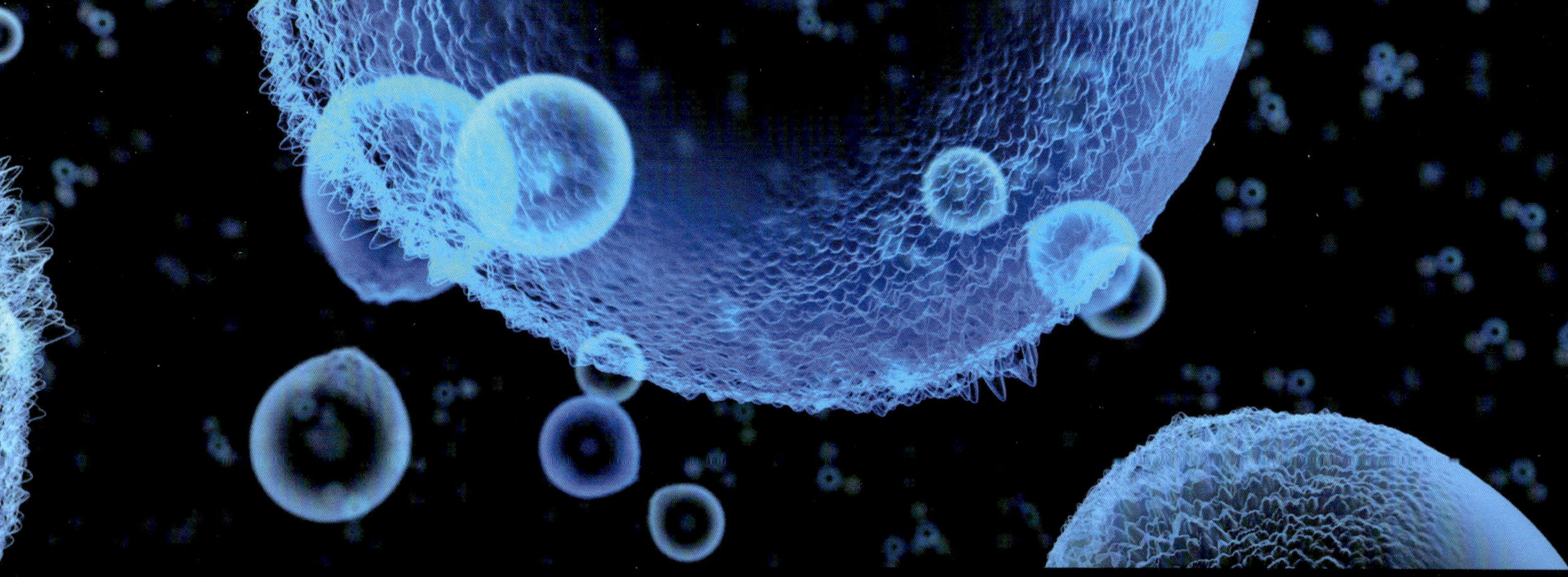

three anti-retroviral drugs, at a fraction of the international cost.

In 1985, Dr Rama Rao turned his attention to immunosuppressant drugs like cyclosporine which are regularly given to organ transplant patients and are very expensive. This was a new area and required asymmetric synthesis. After being successful in this endeavour, he went on to synthesise another compound a hundred times more powerful than cyclosporine which was given to heart transplant patients. With the development of each new process the people of the country gained enormously by having an affordable treatment alternative.

Vancomycin, the last weapon against microbes when all other antibiotics fail, caught Rama Rao's fancy. An extremely complex and fascinating molecule, both in terms of structure and activity, proved to be elusive to anyone who cared to synthesise it, including a professor from Harvard. Ignoring every naysayer, he went ahead and got stuck where everyone else did. His eureka moments have often been during his morning walks. A similar eureka moment found him a solution quite away from where everyone else was looking for it, and he became the first one to complete the synthesis.

Much of Rama Rao's clever and novel ways of synthesis were possible at the time when India followed the process patent, which in a way led to the golden era of Indian pharma industry. With India joining the WTO, the product patent came into vogue, thus denying us an intelligent but shorter method of making drugs affordable to us and the developing world. Adapting to change is another forte of an innovator. Rama Rao continues to innovate in the changed circumstances, and in a matter of fifteen years reinvented himself and scaled new heights.

He has been able to inspire and share his excitement and zest with all those who have worked with him. How else would he have guided 109 PhD students in fifteen years? All through life, each time he was pushed to a corner by lack of resources or an apparently insurmountable challenge, he gave his best and survived the challenge, whether it was his early days in Guntur, his escape into a world of opportunities, or his not finding equipment or basic facilities all through his working life. A harsh exterior but a champion of the underdog is what reflects in his life's work and the man that he is.

None who had seen him grow up in Guntur, living with a maternal aunt on one meal a day, would have ever predicted his future, much less his father who was always worried about his low grades in school and would be more than pleased to see him settled in a low government job. His teachers would have only been pleased to write off this bold boy who was a student leader. It was again on account of Rama Rao's boldness, an element of irreverence, adventure and ability to take risks, apart from the fact that life had already taught him how to the take the rough and tumble with ease, that he found his way to UDCT in Bombay, now known as Mumbai University Institute of Chemical Technology, and the world did not lose an excellent organic and synthetic chemist to a lower grade clerk in a post office, an exam that his father rejoiced when he passed.

—Chandana Chakrabarti

(pages 152-153) India is home to rich oral and written tradition in literature. From the Vedas and Puranas to epics like the *Ramayana* and the *Mahabharata*, Indian literary works have won accolades across the world. In modern times, Indian writers and poets have taken the literary world by storm, and made their mark by winning prestigious honours like the Booker and the Pulitzer.

Amul—the Face of India's White Revolution

Amul, meaning priceless in Sanskrit is the brand of the Gujarat Cooperative Milk Marketing Federation Ltd (GCMMF) which is jointly owned by 2.8 million milk producers making it one of the largest cooperative worldwide. The face of India's White Revolution, it was started in 1946 as a dairy cooperative in a small town of Gujarat. The idea was to protect the milk producers from being exploited by the buyers who would purchase the milk at throwaway prices as it was a perishable commodity. Since, India was still under British rule, the poor farmers and milk producers didn't have much choice therefore they decided to register themselves as a cooperative. With the help of national leaders like Sardar Vallabhbhai Patel and Morarji Desai, they were able to form the Kaira District Cooperative. This helped in changing the way the business of milk was conducted—the collection process was decentralised and village cooperatives were developed to organise the smaller milk producers in the villages.

Dr Verghese Kurien, also known as the Milkman of India, soon joined the cooperative and changed its face forever. Not only was he responsible for broadening the horizon of the cooperative but also for bringing about a change in the way the milk producers worked all over the country. Today, it is the largest producer of milk and milk-based products in India. Amul has a varied range of products that are available in India and abroad. Going way beyond just milk it is now selling pasteurised butter, cheese, ice creams, chocolates, clarified butter, sweets and other products, including probiotic milk and sugar-free ice cream. With the help of a dedicated team and millions of milk producers, he was able to change Amul from a dairy cooperative of Anand to an international name that has given the world 'Amul Pattern' of working.

Under Dr Verghese Kurien, 'Operation Flood' replicated this pattern all over the country joining village level cooperatives to districts which in turn joined the state-level federations. Therefore, it's the primary milk producers living in the villages who govern the structure. This has become a role model for developing nations. Even the World Bank has announced for replication of the Amul model of dairy development in Africa. High level delegations from Tanzania, Kenya, Ethiopia and Uganda expressed a desire to replicate the model in their respective countries.

Technology and innovation plays an important part in this successful model. Chilling units are installed in the villages. Amul helps in the development of embryo transfer and cattle breeding to improve the quality of cattle which in turn improves the milk yields. The dairy cooperative was the first FMCG in India to use the internet to implement B2C commerce giving the customers the option of ordering online. The cooperative is also working towards providing farmers access to information related to markets, technologies and dairy practices through net enable kiosks set up in the villages. The cooperative has also implemented a Geographical Information System at both the ends—from milk collection to the market which helps in keeping track of the entire process.

Amul, an initiative meant for the betterment of the poor milk producers has grown into a US$1.7 billion enterprise which has 15,300 village societies as members, covering 2.9 million milk producers. And the turnover for 2010-11 is expected to be more than US$2.2 billion. The surplus profits earned are ploughed back to farmers as they are the ones responsible for this success story. This capital not only benefits the farmer but also contributes to the development of the village community. Today Amul symbolises many things: high-quality products, reasonable prices, vast cooperative network, power of indigenous technology and the hard work and determination of the farmers' organisation.

Aadhaar—the Unique Identity Card

For a country of more than a billion, to grow is not as big a challenge as to growing inclusively. The divide between the haves and the have-nots is further accentuated because of the identity divide between the two. This is precisely the genesis of the Unique Identification Authority of India (UIDAI) which was established in February 2009 under the chairmanship of Nandan Nilekani, the Co-Founder of Infosys, an acclaimed Indian IT company. The authority's aim is to provide the Unique Identity Card (UID) to all the citizens of the country, including infants and NRIs. The project is part of the Social Security Mission of the government. The endeavour is to include the marginalised, who normally do not have any identification document. The project will also ensure that all the welfare schemes of the government do not remain on paper only and the benefits should reach the 'Bottom of the Pyramid'. This project will also help the government to launch new schemes for the benefit of the society. For instance, this card has been issued to more than six lakh children (from three to eighteen-years-old) in Thiruvananthapuram district of Kerala, with the aim of creating a database of all the children in the state which will help in planning social and academic projects in the future.

Named Aadhaar, the twelve-digit card aims to launch a foolproof identification system in the country, making India the first nation in the world to use biometrics technology in issuing identity cards to its citizens. After filling up the enrollment form, the citizens will have to undergo biometrics test at the UID card centre in their vicinity, where their fingerprints will be registered and iris scanned. To ensure that the work progresses smoothly, their photographs will be clicked on the spot itself. After fifteen days the card will be delivered at the individual's home. There are special methods of enrolling the infants as the biometrics registration is applicable only for children aged above five. In such cases the child's name is to be enrolled with either parent along with his photograph.

In a recent interaction at a National Association of Software and Services Companies (NASSCOM) summit Nandan Nilekani elaborated, 'We are providing identity verification as a cloud service. You can then build applications and embed the verification into your applications. The authentication of an Aadhaar number over a mobile phone would open up opportunities for financial inclusion services.'

'This kind of authentication allows you to do transactions from your mobile,' he said. 'This is going to change the way of banking fundamentally.' The Aadhaar project aims at enrolling about 600 million people in four years through state governments and public sector undertakings such as oil companies.

'Initially, we are doing enrollment through state governments, banks and oil companies but over time, once more people have the Aadhaar numbers, then people are going to build value chain applications. Innovation is going to happen,' Nilekani said. 'We are now examining how to build those kind of applications.'

The stage is set for a breakthrough idea to enable the citizens of the largest democracy to fulfil their dreams and aspirations in more ways than one.

SEWA—Green Energy Finance

There is an overwhelming scientific consensus across the world that global warming is indeed happening and that it is a result of irresponsible human behaviour. This firmly places it and climate change issues as the greatest threat to this planet. We are already seeing its destructive effects in the form of natural disasters—heat waves, tsunamis and

hurricanes—resulting in the death of thousands of people. The tsunami of 2011 that almost destroyed Japan is an unfortunate example of the horrifying effects it can have.

On a smaller, day-to-day level, its effects on the livelihood of the rural household are equally worrying. In fact they are taking place so slowly that most of the times they get left out in the bigger plans and it's only a few groups and organisations that work towards finding a solution. Self-Employed Women's Association (SEWA) is one such group. SEWA, founded by Ela Bhatt, was registered as a trade union in 1972 and is an organisation of the thousands of self-employed women who earn a living by running a small enterprise or by working as daily labours. The aim is to bring them together as an organised group and making them self-reliant. There are over 1,300,000 members across seven states in India and the group is working towards providing microfinance, insurance products, training, rural production and marketing, and housing services to them.

Over the last few years, the rainfall pattern has changed. The incidence of floods has increased thus increasing the area hit by floods. Subsequently, this has resulted in large-scale damage to crops and houses, change in cropping patterns, erosion of fertile soil, increase in types and incidence of various health issues and increased salinity and fluoride content in groundwater.

Invariably, it is the poor farmer who is affected the most by the effects of climate change. The SEWA with its vast sisterhood is acutely aware of this situation and understands all too well that urgent action is required to avert the impending catastrophe. A concerted action is, however, not forthcoming from the developed nations of the world, and efforts by the world's governments to address climate change is missing in speed and coordination. Fresh thinking and innovative approaches by non-government agencies are hence the need of the hour.

The Social Impact Finance (SIF) is one such approach promoted and developed by a number of agencies and individuals in an attempt to address paradoxical situations where prevention of social and health problems clearly saves public money, but finding the funds and incentives to do so is difficult for the public bodies.

The mechanism of SIF is as follows:

- Step One: Identify a project that would create a certain social impact, and to estimate the total fund required for the implementation.
- Step Two: Raise funds for this intervention from investors, and in turn issue them finances commensurate with their investment. These finances would have a capped rate of return and cash flows at certain designated intervals.

- Step Three: Identify an agency (underwriter) that would guarantee this return to the investors. Typically this would be the public sector. The underwriting agency would decide a return in proportion to the measured social impact. Hence the risk of this finance lies with the investors, and is in their interest to ensure the best possible impact for their investment.

SEWA takes inspiration from the SIF and believes that a project that reduces the carbon footprint could be ideally financed and implemented through a similar scheme—a finance that it labels the Green Energy Finance or the Hariyali Finance.

The social impact, investment and underwriting element of the Green Energy Finance will be seen in years to come. SEWA plans to provide 2,00,000 poor women living in India, Sri Lanka, Afghanistan and Nepal with access to safe, economical and environment-friendly cooking stoves and solar lights in three years. Access to these women will be via the organisation's existing rural distribution network. SEWA will work with leading manufacturers to ensure that the design of the cooking stoves and solar lights meet the specific needs of the women.

The project will have a significant impact in the following areas:

- Improved health through reduction in illnesses associated with indoor air pollution (eg. loss of eyesight, respiratory problems including TB, cancer).
- Increased income generation through time and financial savings associated with sourcing and acquiring fuel and reduced ill-health.
- Improved school attendance and performance for children through reduction in illnesses, ability to study at home and increased income to fund studies.
- Youth employment and training through the involvement of up to 2,500 youngsters in the (distribution and maintenance) of the cooking stoves and solar lights.

(facing page) An Indian training supervisor (L) trains a young Pakistani woman as she uses a sewing machine at SEWA Trade Facilitation Centre (STFC) in Ahmedabad, Gujarat. In 2009, a batch of some twenty women from the remotest and poorest regions of Pakistan undertook training as part of SAARC'S project, SABAH.

A model walks the ramp at the launch of a collection, 'Ananta', by STFC at the National Crafts Museum in New Delhi. Established in 2003, with headquarters in Ahmedabad, STFC's aim is to provide socio-economic security to craftswomen at the grass-roots level.

The project cost is estimated at ₹50 crores, which would be the Green Energy Finance's nominal value. This is to be issued to investors in ten or more tranches of up to ₹5 crores each. The Finance's duration will be five years. Coupon payments and repayment terms will be worked out at the next stage. The following avenues of revenue generation would make up the return for the investors:

- It is estimated that the cooking stoves and solar lights will have a respective maximum cost of $22 and $30 per piece and will be sold at the base cost price. A part of this would be payable upfront, the rest would be structured as a loan payable as instalments over a designated period.
- A government guarantee is being sought from the Dutch, Swedish, Norwegian and Indian governments whereby they agree to underwrite the financial return, subject to the level of social impact achieved.
- Revenue generation through carbon credits would be explored as well.
- SEWA seeks the participation of interested investors in this project. A successful implementation of this project would set the stage for the replication of this model for the implementation of projects with different causes and a larger scale.

Photograph Credits